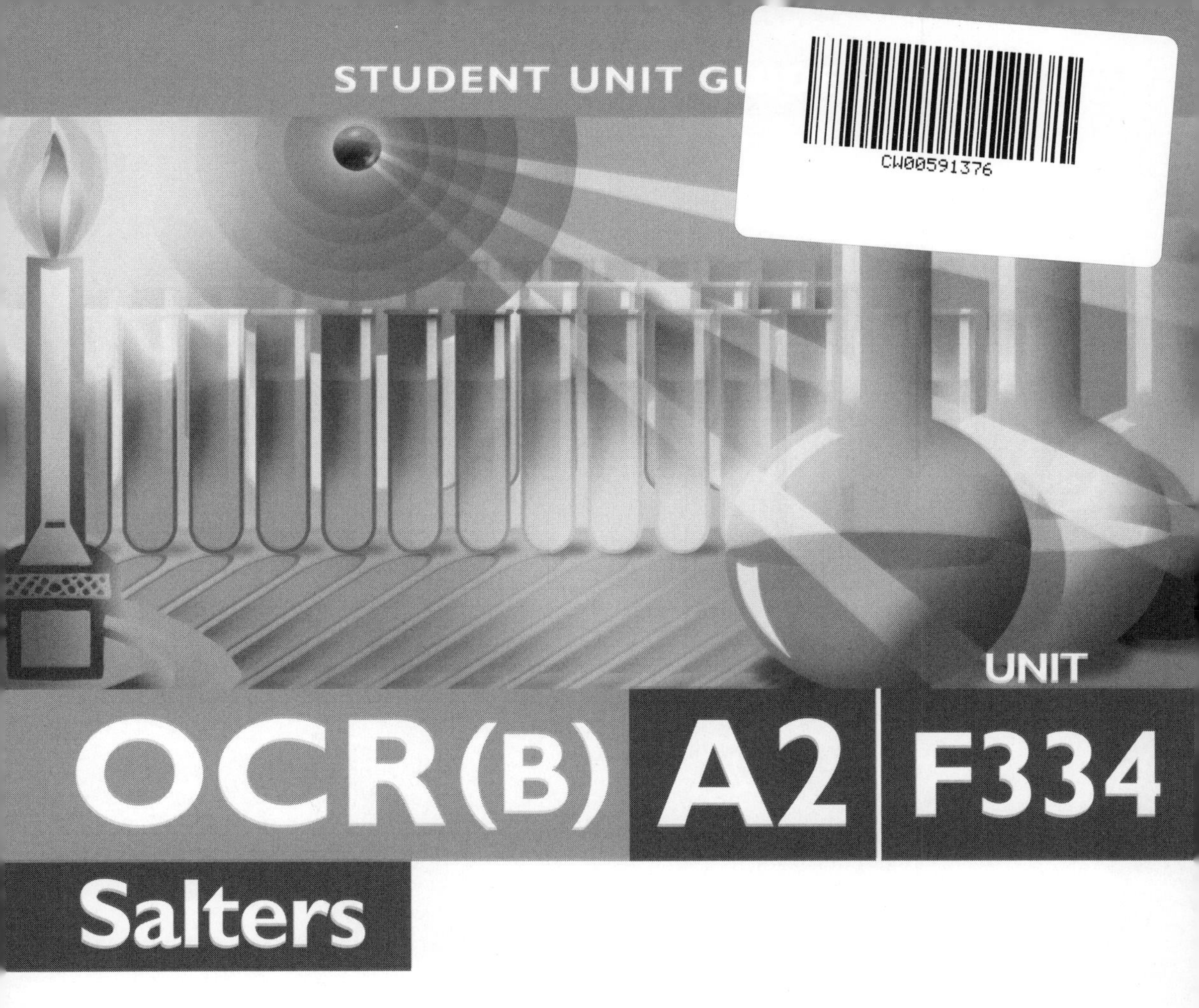

UNIT

OCR (B) A2 | F334

Salters

Chemistry

Chemistry of Materials

Frank Harriss

To Maggi

Philip Allan Updates, an imprint of Hodder Education, an Hachette UK company, Market Place, Deddington, Oxfordshire OX15 0SE

Orders

Bookpoint Ltd, 130 Milton Park, Abingdon, Oxfordshire OX14 4SB
tel: 01235 827720
fax: 01235 400454
e-mail: uk.orders@bookpoint.co.uk

Lines are open 9.00 a.m.–5.00 p.m., Monday to Saturday, with a 24-hour message answering service. You can also order through the Philip Allan Updates website: www.philipallan.co.uk

ISBN 978-0-340-94823-1

First printed 2009
Impression number 5 4 3 2 1
Year 2014 2013 2012 2011 2010 2009

Typeset by Pantek Arts Ltd, Maidstone

Printed by MPG Books, Bodmin

Hachette UK's policy is to use papers that are natural, renewable and recyclable products and made from wood grown in sustainable forests. The logging and manufacturing processes are expected to conform to the environmental regulations of the country of origin.

P01412

Contents

Introduction

About this guide

This book is designed to help you prepare for the first OCR (Salters) A2 Chemistry B Specification unit test, which examines the content of **Unit F334: Chemistry of Materials**. This unit covers:

- What's in a medicine?
- The materials revolution
- The thread of life
- The steel story

The aim of the guide is to provide you with a clear understanding of the requirements of the unit and to advise you on how best to meet them.

The book is divided into the following sections:

- This **Introduction**, which outlines revision and examination technique, showing you how to prepare for the unit test.
- **Content Guidance**, which provides a summary of all the new 'chemical ideas' in Unit F334, together with important revision ideas from previous units.
- **Questions and Answers**, in which you will find questions in the same style as in the unit test, followed by the answers of two students, one of whom is likely to get an A grade, the other a C/D grade. Examiner's comments follow these answers.

How to use this guide

- Read the section 'Revision and examination technique' in this introduction.
- Decide on the amount of time you have available for chemistry revision.
- Allocate suitable amounts of time to:
  - each section of the Content Guidance giving the most time to the areas that seem most unfamiliar
  - the questions from the Questions and Answers section
- Draw up a revision timetable, allocating the time for questions later in your timetable.
- When revising sections of the Content Guidance:
  - read the guidance and look at corresponding sections in your notes and textbooks
  - write your own revision notes
  - practise answering questions from past unit tests and other sources, such as *Chemical Ideas*.
- When using the Questions and Answers:
  - try to answer the question yourself

- then look at the students' answers, together with your own and try to work out the best answer
- then look at the examiner's comments

Revision and examination technique

Do I have to remember material from the AS units?

This is a 'synoptic' unit, which means that it tests all the chemistry you have studied so far. So the answer to the question is 'yes'. To make it easier for you, there are synoptic statements in the specification for the unit which indicate where ideas from AS will be revisited.

How do I find what to learn?

In addition to this guide, other useful sources are:

- the specification. This is the definitive source. If it's not in the specification, it won't be in the exam paper. If you need amplification of a specification statement, look in the *Chemical Ideas* for the depth of treatment. This guide should help you to interpret the content, as every new specification point for the Salters topics in this unit is covered in the Content Guidance section.
- the 'Check your knowledge and understanding' activities in the *Activities* pack. They suggest sources of details not found in the *Chemical Ideas* book. Some of the material is in the *Storylines* book and some in the activity sheets.
- your own and your teacher's notes. Preparation for an exam is not just something you do shortly before you take the paper. It should be an integral part of your daily work in chemistry.

How much of the *Storylines* and *Activities* do I need to learn?

Have a look through for yourself, but you will find the details in the 'Check your knowledge and understanding' activity sheets referred to above.

The primary function of the *Storylines* book is to provide a framework and a justification for studying the theory topics. For all four Salters topics in this unit there are important parts of the theory in the *Storylines*.

The *Activities* are provided to teach practical and other skills and to back up the theoretical ideas. However, they may also contain some theory that does not occur elsewhere.

General revision tips

Revision is a personal thing

What works for one person does not work for another. You should by now know which methods suit you but here are a few ways to set out your revision notes.

- Mind maps — ideas radiate out from a central point and are linked together. Some people like to colour these in.
- Notes with bullet points and headings.
- Small cards with a limited 'bite-size' amount of material on each.

Make a plan

Divide up your material into sections. It is probably best to revise by Chemical topics at this stage, rather than Salters topics. (The Content Guidance section will be helpful here.)

- Work out how much time you have available before the exam.
- Allocate each section as much time as you can, bearing in mind which you feel you nearly understand and which are the most difficult.
- Fit this in with any revision your teacher is going to do — ask him or her for a summary.

Write, write, write

Whatever you do, make sure that your revision is *active*, not just flipping over the pages, thinking that you know it already. Write more revision notes, test yourself (or each other) and try questions.

Test yourself

- The questions in *Chemical Ideas* are useful 'drill exercises' on topics, but they are not all like exam questions.
- If you have taken end-of-module tests, go through them again and then check your answers against the corrected version or the model answers you may have been given. They are much more like exam questions.
- Past papers (those for Module 2849 of the old specification will be useful) give you a good indication of what you will be facing.
- The Question and Answer section of this book is designed specifically to allow you to test yourself.

Types of knowledge

- **Recall.** This includes definitions, diagrams and methods, for example organic reactions. The specification statements will often start 'recall', 'recognise' or 'describe'. Learn this material thoroughly when you first meet it. Revise and summarise it later; then you can look through it shortly before the examination.
- **Understanding.** This requires knowledge of methods and basic factual information but you will need practice in applying the ideas you know to these situations. Calculations and explanations come into this category, also specification statements beginning 'explain' and 'understand'. The practice should be done well before the examination so that you have a chance to sort out the ideas.

The day or two before the exam

- **Should you revise?** Yes. But make sure it's recalling the facts rather than trying to understand concepts — this should have been done earlier.
- **Read through the *Storylines*.** Some of the questions may be set in the context of the *Storylines* and this will help you put your knowledge into perspective.
- ***Mens sana in corpore sano*** (A sound mind in a healthy body). Do take some exercise or play sport. You will feel much better for this, rather than sitting the whole time in a darkened room revising.
- **Check the administrative details.** When does the paper start? How long is the paper? How many marks? What materials do I need? Which teaching modules are being examined?
- **A good night's sleep.** Aim to have several of these the nights before the exam. Then if you can't get to sleep the night before, it matters far less.

Know the enemy — the exam paper

I hope, since you will have prepared properly, you will be able to look on the exam as an opportunity to show what you can do, rather than as a battle. Be aware, however, that you must prepare yourself for an exam just as you would for an important sporting contest — be focused. Work hard right through the 90 minutes and do not dwell on difficulties — put them behind you. Try to emerge feeling worn out but happy that you have done your best, even if you have found it difficult (others will probably feel the same way). Then forget it and don't have a post-mortem.

Every question tells a story

Salters is all about learning chemistry in relevant (i.e. real-life) contexts, so it is right that the exam questions should reflect this. Sometimes the context will come from the *Storylines*; sometimes it will be a new one. Look carefully at the 'stem' (the introduction at the top of the question). Most of the important facts here will be needed somewhere in the question. Often there are small additional stems later on. These are important too.

90 marks in 90 minutes

It is important to 'pace yourself' through the paper so you can tell whether you are ahead of, or behind, the clock. It is best to work through the paper in order, from the beginning to the end, since the first question is intended to be one of the easier ones. There are usually four or five questions.

Knowledge or application of knowledge

About half the marks will test your knowledge and ask about things you will have learned. The other half of the marks are for the application of that knowledge to new situations or through doing calculations. These questions sometimes begin

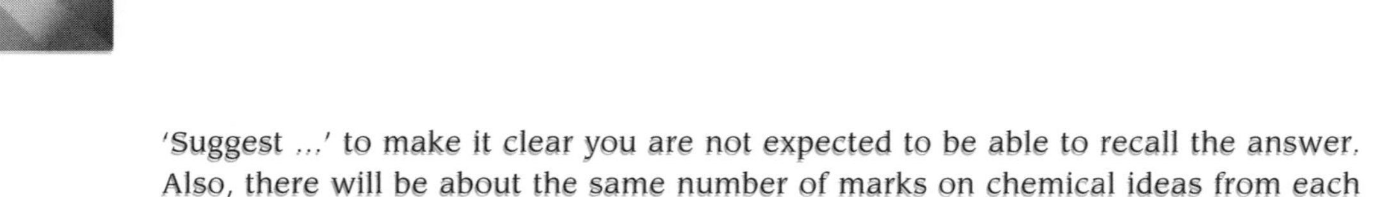

'Suggest ...' to make it clear you are not expected to be able to recall the answer. Also, there will be about the same number of marks on chemical ideas from each teaching module.

Easy and hard parts

The papers are designed so that, ideally, an A-grade candidate will get 80% (72 out of 90) and an E-grade candidate 40% (36 out of 90). The actual mark for each grade varies between papers, depending on the difficulty and is only decided after all the papers have been marked. Some of the marks are designed with A-grade and A*-grade candidates in mind and thus will seem quite demanding. Other marks are designed in order to allow an E-grade candidate to score 40% and thus will seem rather easy. Thus there are easy, middling and hard parts within each question.

Dealing with different types of question

Short-answer questions

These are the most straightforward, but remember:

- Look at the marks available — make one good point per mark.
- Look the number of lines — this gives *some* idea of the length of answer required. Handwriting differs greatly in size, but if you have written two words and there are three lines, you may assume you have not written enough to score full marks.
- Don't 'hedge your bets' — if you give two alternative answers, you will not get the marks unless *both* are right. For example, if the answer is 'hydrogen bonds' and you write 'hydrogen bonds or permanent dipole–permanent dipole interactions', you will score zero.
- Read the question — don't answer a question that you have made up. Examiners do have kind hearts really and they are genuinely sorry when they have to award zero for an answer containing good chemistry that is not relevant to the question asked. This is a problem with units that are examined twice a year. There are many past papers for the old Module 2849, which is similar to F334. They ask similar but slightly different questions on a topic. It's all too easy to give the answer to last year's question.

Long-answer questions

In this unit test, about 10% of the marks are for extended answers. The same rules apply about marks and lines and reading the question.

- Think before you write — perhaps jot a few points in the margin. Try to make your points logically.
- Punch those points — if you read any mark schemes, you will see that they give examiners advice on the weakest answer that will still just score the mark. Make sure your points are well made and win the mark without a second's hesitation by the examiner.

- Try to write clear sentences (though bullet points might be appropriate on some occasions).
- Be sure not to re-state the question, i.e. don't use words or phrases directly from the question as part of your explanation.

Quality of written communication (QWC)

In each F334 paper, you will see the ✎ symbol twice.

On one occasion it will be accompanied by the words:

✎ *'In your answer you should use appropriate technical terms, spelt correctly.'*

In such answers, you should ensure that you do use technical terms, rather than less focused alternatives. For example, use 'absorbance' rather than 'reading' from a colorimeter. You must also ensure, obviously, that you spell these words correctly.

The wording for the other ✎ instruction starts:

✎ *in your answer you should make it clear...*

and then continues with wording such as:

...how your explanation links with the chemical theory/evidence

...how the steps you describe are linked to one another

...how the structures you describe are related to their properties

Here you should think carefully about the logic of your answer and not leave out linking steps. For example, if you say 'weak intermolecular bonds mean greater flexibility of the polymer', you might lose the 'QWC' mark for not making the linking point 'the chains can move over each other more easily'.

Command words in questions

A lot of care is taken in choosing which commands words to use, so note them carefully:

- 'State', 'write down', 'give' and 'name' require short answers only.
- 'Describe' requires an accurate account of the main points but no explanation.
- 'Explain' requires chemical reasons for the statement given.
- 'Suggest' means that you are not expected to know the answer but you should be able to work it out from what you do know.
- 'Giving reason(s)' requires you to explain why you chose to answer as you did (if 'reasons' in the plural is stated, judge the number required from the number of marks).

Avoid vague answers

Sometimes it is clear that the candidate knows a lot about the topic but his or her answer is not focused. Avoid these words:

- 'It' (e.g. 'it is bigger') — give the name of the thing you are describing. Otherwise it may not be clear which object in the question is being referred to.
- 'Harmful' — if you mean 'toxic' or 'poisonous', say so.
- 'Environmentally friendly' — say *why* it benefits the environment.
- 'Expensive' — always justify this word with a reason.
- Be careful with chemical particles — always think twice whenever you write *particle, atom, molecule* or *ion* and check that you are using the correct term.

If in doubt, write something

Try to avoid ever leaving gaps. Have a go at every answer. If you're not sure, write something that seems to be sensible chemistry. As you will see from the Question and Answer section, some questions have a variety of possible answers — the only answer that definitely scores zero is a blank.

Diagrams

You would be amazed at some of the diagrams examiners have to mark, so please:

- **Read the question.** The answer is not always a reflux condenser. If it is an apparatus you know, then it is relatively straightforward. If you have to design something, look for clues in the question.
- **Make it clear and neat.** Use a pencil and ruler and have a soft rubber handy to erase any errors. (See 'on-screen marking' on p. 11.)
- **Make sure it looks like real apparatus** (which never has square corners, for example). Some apparatus drawn in exams would test the skill of the most proficient glass-blower.
- **Draw a cross-section so that gases can have a clear path through.** Don't carelessly leave any gaps where gases could 'leak out'.
- **Think of safety.** Don't suggest heating an enclosed apparatus, which would explode. If a poisonous gas is given off, show it being released in a fume cupboard.
- **Always label your diagram,** especially if the question tells you to. Important things to label are substances and calibrated vessels (e.g. syringes or measuring cylinders).

Calculations

If you get the answer to a calculation right, the working does not need to be there (unless you could have guessed the answer). However, it is always easy to make mistakes, especially under the pressure of exams. So, set out the steps in your calculations clearly. Then you will get most of the marks if you make a slight mistake and the examiner can see what you are doing. Examiners operate a system called 'error carried forward' whereby, once an error has been made, the rest of the calculation scores marks if the method is correct from then on.

At the end of the calculation, there will be a line that reads for example:

Volume =________________ (2 marks)

Obviously you should write your answer clearly here. When you write down your numerical answer, check:

- **units** — most physical quantities have them. (Sometimes these appear on the answer line to help you, but where they don't, you must supply them.)
- **sign** — remember that oxidation states and ΔH values must be shown as '+' if they are positive.
- **significant figures** — you may be expected to analyse uncertainties more carefully in your practical work, but in exam papers all you have to do is to give the same number of significant figures as the data in the question.

On-screen marking

When you have written your answers, the papers are collected and sent to a scanning centre. There they are scanned so that they can be marked online by examiners. This has the advantage to you that examiners' marking can be constantly monitored and, if they have a query on your script, they can easily refer it to a senior examiner. Also, there is less chance of a clerical error at any stage.

This process should not affect you, but it is worth noting the following points:

- Use black ink (use pencil only for diagrams and graphs).
- Do not use colours, as these will only come out in shades of grey and may not be clear.
- Write clearly, as scanned writing can be more difficult to read. If you are unsure, check by photocopying some of your work.
- Rub out any incorrect pencil marks fully as the scanning is sensitive and the examiner cannot always see which is the intended answer. If you wish to make several changes to a diagram and there is room, it may be best to cross it out and start again.
- Do not 'over-write' letters or numbers, for example, changing 'ethanal' to 'ethanol' by writing over the 'a'. Cross out the whole word and start again. There are no marks for neatness — just ensure it is clear exactly what you mean.
- Do not continue an answer below the dotted lines unless it is the last question on the page.
- Ask for extension sheets rather than using other blank spaces on the paper.

Be assured, however, that examiners will do their utmost to give you the credit your answers deserve, even if they are written in the 'wrong' places.

Content Guidance

The material in this section summarises the chemical ideas from **Unit F334: Chemistry of Materials**. It is arranged in a logical chemical order, not the order in which you study it (which is determined by the content of *Storylines*). Revision of AS material is shown in *italics*.

Summary of content

Organic chemistry: functional groups and reaction types (*part revision*); carboxylic acids, esters, phenols and acyl chlorides; aldehydes and ketones; amines, amides; stereoisomerism; organic techniques

Condensation polymers: condensation reactions; polyesters; polyamides; hydrolysis; properties of polymers; varying the properties

Proteins: amino acids; protein structure; enzymes

DNA: structure; protein synthesis; DNA databases

Reaction rates: the effect of concentration; enzyme kinetics

Transition metals: electronic structure; typical properties — variable oxidation state, formation of complexes, formation of coloured ions, catalytic behaviour;

Redox: *revision*; half-equations; cells and electrode potentials; use of electrode potentials; rusting

Analytical methods: colorimetry; manganate(VII) titrations and mole calculations; mass spectrometry; infrared spectroscopy (*revision*)

Industrial and green chemistry: extraction and purification of metals; making and testing medicines; atom economy and reaction type; the polymer 'life cycle'; recycling steel; enzymes in industry

How much of this do I need to learn?

The answer is virtually all of it. It has been pared down to the absolute essentials. If you need any more detail on any aspect, you should look in your textbooks or your notes.

Organic chemistry

Functional groups and reaction types

The first three functional groups are revision of AS chemistry. You will not have met the others in detail, so there are notes on them below the table.

Tip You need to know about all the information in the table, apart from the examples, which you must understand.

Type of compound	Functional group	Name	Type of reaction	Example	Test
Alkene	C=C	-ene	Electrophilic addition	$C_2H_4 + HBr \rightarrow C_2H_5Br$	Bromine water turned from brown to colourless in cold.
Halogenoalkane	–Br, –Cl, –I	chloro- etc.	Nucleophilic substitution	$C_2H_5Br + NaOH \rightarrow C_2H_5OH + NaBr$	
Alcohol	–OH (joined to alkyl group)	-ol	Nucleophilic substitution	$C_2H_5OH + HBr \rightarrow C_2H_5Br + H_2O$	Primary and secondary alcohols turn acidified dichromate from orange to green (oxidation reaction).
			Dehydration (elimination)	$C_2H_5OH \rightarrow C_2H_4 + H_2O$	
			Oxidation	$C_2H_5OH \rightarrow CH_3CHO \rightarrow CH_3COOH$	
			Condensation (esterification)	$C_2H_5OH + CH_3COOH \rightarrow CH_3COOC_2H_5 + H_2O$	
Phenol (see p. 16)	O–H attached to benzene ring	–	Acts as an acid	$C_6H_5OH + NaOH \rightarrow C_6H_5O^-Na^+ + H_2O$	Purple colour with iron(III) chloride
			Condensation (esterification)	$C_6H_5OH + CH_3COCl \rightarrow CH_3COOC_6H_5 + HCl$	
Carboxylic acid (see p. 16)	–C(=O)–O–H	-oic acid	Acidic	$2CH_3COOH + Na_2CO_3 \rightarrow 2CH_3COO^-Na^+ + CO_2 + H_2O$	Fizz with sodium carbonate solution
			Condensation (esterification)	$C_2H_5OH + CH_3COOH \rightarrow CH_3COOC_2H_5 + H_2O$	

Type of compound	Functional group	Name	Type of reaction	Example	Test
Acyl chloride (see p. 16)	O C Cl	(-oyl chloride)	Condensation (acylation)	$CH_3COCl + RNH_2 \rightarrow$ $CH_3CONHR + HCl$	
(Primary) amine (see p. 19)	$-NH_2$	Amino-	Acts as base	$RNH_2 + HCl \rightarrow RNH_3^+Cl^-$	
			Condensation	$CH_3COCl + RNH_2 \rightarrow$ $CH_3CONHR + HCl$	
Ester (see p. 16)	O C O	See p. 18	Hydrolysis	$CH_3COOC_2H_5 + NaOH \rightarrow$ $CH_3COO^-Na^+ + C_2H_5OH$	(Sweet smell)
Amide (see p. 20)	O C N H	–	Hydrolysis	$CH_3CONH_2 + NaOH \rightarrow$ $CH_3COO^-Na^+ + NH_3$	
Aldehyde (see p. 18)	O C H	-al	Oxidation	CH_3CHO to CH_3COOH (acid dichromate, reflux)	
			Nucleophilic addition	$CH_3CHO + HCN \rightarrow$ $CH_3CH(OH)CN$ (cyanohydrin)	
Ketone (see p. 18)	O C	-one	Nucleophilic addition	$CH_3COCH_3 + HCN \rightarrow$ $CH_3C(OH)(CN)CH_3$ (cyanohydrin)	

Carboxylic acids, esters, phenols and acyl chlorides

You meet these compounds in **What's in a medicine?**

Carboxylic acids contain the group:

O C O — H

The –COOH group is acidic as it loses a proton in aqueous solution:

$$CH_3COOH \rightleftharpoons CH_3COO^- + H^+$$

An acid is defined as a proton donor (by the Brønsted–Lowry theory)

Phenols are also weakly acidic:

$$C_6H_5OH \rightleftharpoons C_6H_5O^- + H^+$$

Thus, both react with alkalis such as sodium hydroxide:

$$CH_3COOH + NaOH \rightarrow CH_3COONa + H_2O$$

Ethanoic acid — Sodium ethanoate

$$C_6H_5OH + NaOH \rightarrow C_6H_5ONa + H_2O$$

Phenol — Sodium phenoxide

Carboxylic acids react with carbonates to give carbon dioxide, but phenols are not strong enough to do this.

$$2CH_3COOH + Na_2CO_3 \rightarrow 2CH_3COONa + CO_2 + H_2O$$

Phenols give a violet colour with neutral iron(III) chloride. They can be esterified but only by reaction with acyl chlorides or acid anhydrides. They do not react with carboxylic acids.

Acyl chlorides can be made from carboxylic acids (by a reaction you do not need to remember — reaction 6 on the *Data Sheets*). They contain the group:

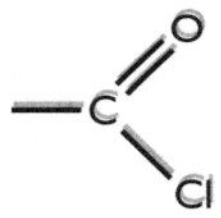

$$—C(=O)Cl$$

The presence of the C=O makes the C–Cl bond very reactive. Thus, while carboxylic acids have to be heated with alcohols to make esters, acyl chlorides only need to be mixed with alcohols at room temperature.

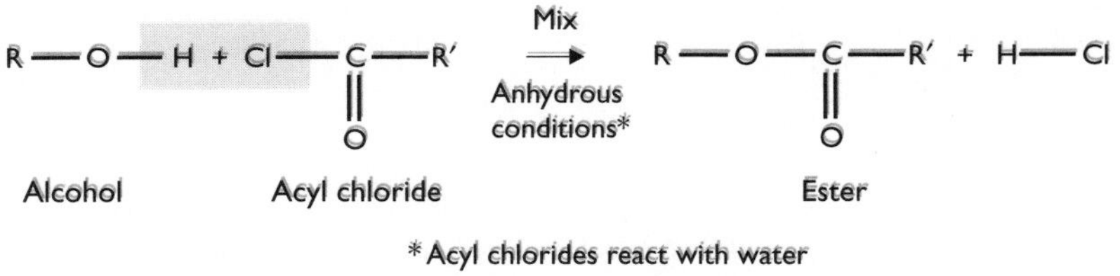

$$R—O—H + Cl—C(=O)—R' \xrightarrow[\text{Anhydrous conditions*}]{\text{Mix}} R—O—C(=O)—R' + H—Cl$$

Alcohol — Acyl chloride — Ester

* Acyl chlorides react with water

Acyl chlorides also condense with amines when they are mixed at room temperature — see below.

Acid anhydrides can also be used in rapid esterification. They are made by removing one water molecule from two carboxylic acid molecules.

$$2CH_3COOH - H_2O \rightarrow (CH_3CO)_2O$$

Naming acids and esters

Carboxylic acids

These are named by counting all the carbon atoms:

e.g.			
CH_4	methane	HCOOH	methanoic acid
C_3H_8	propane	C_2H_5COOH	propanoic acid

Esters

When naming these, remember that they are made from an acid reacting with an alcohol. Identify the alcohol group (attached to the O) and name it (e.g. methyl, ethyl). Then identify the acid group (including the C=O) and name this (e.g. ethanoate, propanoate).

Tip Note that the formulae of esters can be written either way round as these examples show.

Ester	Alcohol (dark area)	Acid (light area)	Name
$CH_3-C(=O)-O-CH_3$	Methanol Thus 'methyl'	Ethanoic acid Thus 'ethanoate'	Methyl ethanoate
$CH_3-CH_2-O-C(=O)-H$	Ethanol Thus 'ethyl'	Methanoic acid Thus 'methanoate'	Ethyl methanoate

Aldehydes and ketones

These compounds are mentioned in **The polymer revolution**, but you meet their reactions for the first time in **What's in a medicine?**

	Aldehyde	Ketone
Formula	$R-C(=O)-H$	$R-C(=O)-R'$
Formation	Oxidation of a *primary*, alcohol by **distilling** with acidified potassium dichromate $CH_3-CH_2-OH \rightarrow CH_3CHO$	Oxidation of a secondary alcohol by **heating under reflux** with acidified potassium dichromate $H_3C-CH(OH)-CH_3 \rightarrow H_3C-C(=O)-CH_3$
Oxidation	Can be further oxidised by **heating under reflux** acidified potassium dichromate with $CH_3CHO \rightarrow CH_3COOH$ A carboxylic acid is formed	Cannot be further oxidised (without breaking up the molecule)

	Aldehyde	Ketone
Reduction This reagent is on the *Data Sheets*, you do not need to learn it	Can be reduced back to a *primary* alcohol using $NaBH_4$	Can be reduced back to a *secondary* alcohol using $NaBH_4$
Reaction with hydrogen cyanide nucleophilic addition (for details, see below)	$CH_3CHO + HCN \rightarrow$ 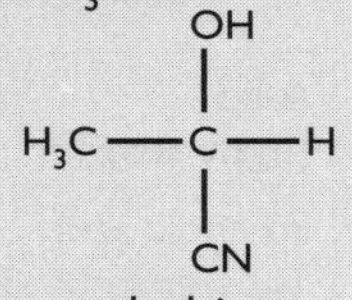 $H_3C-C(OH)(CN)-H$ cyanohydrin	$CH_3COCH_3 + HCN \rightarrow$ 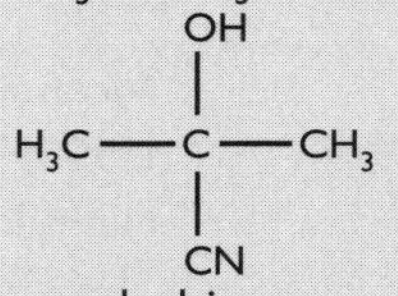 $H_3C-C(OH)(CN)-CH_3$ cyanohydrin

Mechanism of nucleophilic addition

The lone pair of electrons on the cyanide (^-CN) ion attacks the slightly positive carbon of the C=O group and forms a new bond. (Remember that a *curly arrow* indicates the movement of a pair of electrons.)

One pair of electrons from the C=O bond moves to the oxygen atom, making it negatively charged. The negatively charged ion removes a proton from water, the solvent. The overall effect is the addition of HCN to the carbonyl compound.

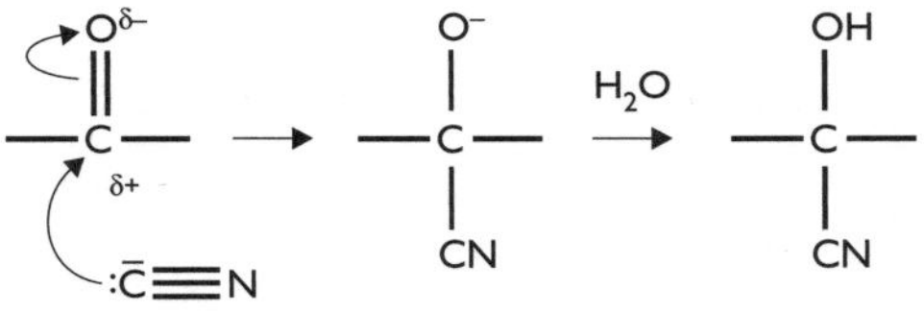

$$>C^{\delta+}=O^{\delta-} + {}^-:C\equiv N \longrightarrow -C(O^-)(CN)- \xrightarrow{H_2O} -C(OH)(CN)-$$

Amines

You meet amines in **The materials revolution**.

Amines are described as **primary**, **secondary**, and **tertiary**. This is *not* the same definition as for alcohols. For *alcohols*, it is the number of alkyl (R) groups on the carbon next to the –OH group that must be counted (one for primary etc.). For *amines* it is the number of alkyl groups on the nitrogen itself that must be counted.

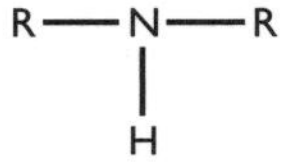

$R-N(H)-R$

Secondary amine

$R-C(H)(R)-OH$

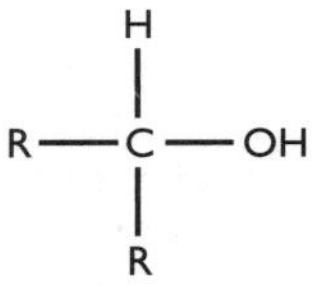

Secondary alcohol

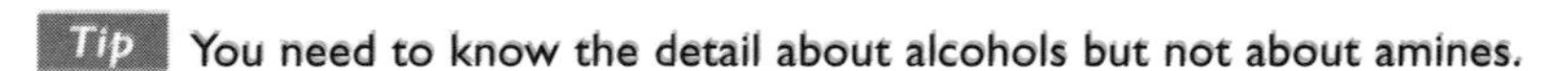

Tip You need to know the detail about alcohols but not about amines.

Primary amines are named thus:

$H_3C—N(H)—H$ is aminomethane

$C_2H_5—N(H)—H$ is aminoethane

Properties of amines

(1) They are basic, forming alkaline solutions in water:

$RNH_2 + H_2O \rightleftharpoons RNH_3^+ + OH^-$

and reacting with acids to accept a proton:

$RNH_2 + HCl \rightarrow RNH_3^+Cl^-$ (an amine salt)

Bases are **proton acceptors** on the Brønsted–Lowry theory.

The lone pair on the nitrogen atom accepts the proton to give a cation.

(2) They undergo **condensation reactions** with acyl chlorides to form amides:

(This is also known as **acylation**.)

$R—N(H)—H + Cl—C(=O)—CH_3 \rightarrow R—N(H)—C(=O)—CH_3 + H—Cl$

Amides

Amides contain this group:

$—N(H)—C(=O)—$

Primary and secondary amides are defined like primary and secondary amines.

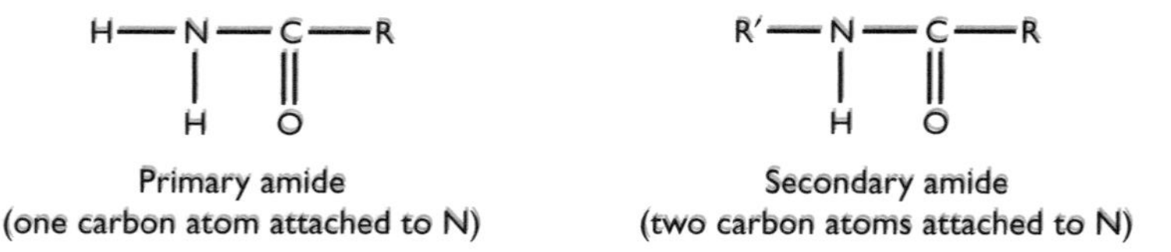

Primary amide (one carbon atom attached to N)

Secondary amide (two carbon atoms attached to N)

Tip You need to be aware that $RCONH_2$ and RCONH– are both amides, but you are not expected to learn any other detail.

Stereoisomerism

In **Developing fuels** you met the idea of structural isomers — molecules with the same molecular formula but a different structural formula. Here we look at molecules that have the same structural formula but differ because of the *arrangements of the atoms in space.*

E/Z isomerism

You met this concept in **The polymer revolution**. *E/Z* isomerism arises when there is a C=C double bond in a molecule and each carbon has two different groups attached. The groups around a C=C bond cannot rotate like those around a C–C bond.

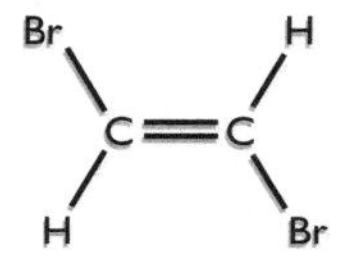

(*E*)-1,2-dibromoethene
or *trans*-1,2-dibromoethene

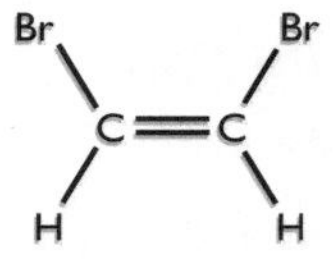

(*Z*)-1,2-dibromoethene
or *cis*-1,2-dibromoethene

Tip You have to know only how to define *E/Z* when the two groups on each carbon atom are the same (here they are Br and H). You must be able to extend the *cis/trans* definition to other substances.

- To *identify E/Z* isomerism — look for a C=C double bond with *two different groups* attached to each side.
- To *explain E/Z* isomerism refer to the *lack of free rotation* about C=C bonds.

Optical isomerism

You meet optical isomerism in **The thread of life**. It occurs when a molecule has a *mirror image* that is different from the original molecule.

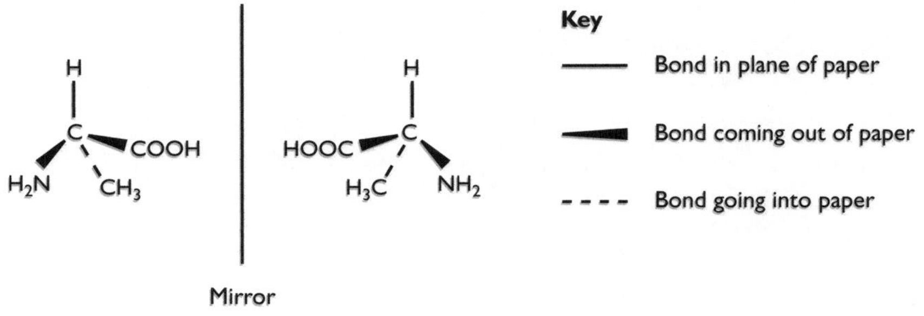

Key
Bond in plane of paper
Bond coming out of paper
Bond going into paper

Mirror

The two *optical isomers* (**enantiomers**) of the amino acid alanine are shown.

These molecules are described as **chiral** (from the Greek word for 'hand' — they are related in the same way as a right and left hand are, as non-superimposable mirror images).

- To identify chiral molecules, look for a carbon atom with four different groups (sometimes called an *asymmetric* carbon atom).
- To draw chiral molecules, use the style as shown in the diagrams above — wedge-shaped bonds, lines and dashed lines. Put in a mirror line and draw both the object and its reflection.
- To explain chirality, you should refer to the object and its mirror image as being **non-superimposable**.

Organic techniques

Purifying solids

You meet this technique in **The materials revolution**.

When an organic solid has been made, it is purified by **recrystallisation**. This involves the steps in the table below:

Step	Explanation
Dissolve in the *minimum* volume of hot solvent (in which the solid is more soluble when hot than when cold).	The solid being purified must just saturate the hot solvent. Impurities, being present in smaller amounts, will not saturate the solvent.
If the solution is cloudy, filter hot.	This removes any *insoluble* impurities.
Allow the solution to cool and crystallise. (This is the *recrystallisation*.) Do not let all the solvent evaporate.	The required solid is less soluble in cold solvent than hot solvent, so it will crystallise. Impurities remain in solution (unless the solvent evaporates).
Filter off the crystals.	The impurities are still in solution.
Wash with a little *cold* solvent.	To wash off the impurities.
Dry (by sucking air through the filter if a volatile solvent has been used).	This removes any remaining solvent.

The purity of a solid can be checked by measuring its melting point. An impure solid always melts below the expected temperature. Pure substances melt sharply, while impure ones melt over a range of temperatures.

Thin-layer chromatography

You meet this topic in **What's in a medicine?**

- Spots of unknown and known substances are applied to a base line on a thin-layer plate.

- The plate is placed in a solvent (so the spots are above the level of the solvent) in a beaker.
- The beaker is covered.
- The solvent rises up the plate. Its final level is marked.
- The plate is taken out, dried and treated with a locating agent (often ultraviolet light or iodine vapour) so the spots show up.
- Spots of the same substance rise to the same characteristic height.
- $R_f \text{ value} = \dfrac{\text{distance moved by spot}}{\text{distance moved by solvent front}}$

Heating under reflux

You met this in **The polymer revolution**. This technique is often used in organic chemistry. It involves mounting a water-cooled condenser vertically above a flask. Its purposes are:

- to allow liquids to be heated to help them react, avoiding the loss of vapours (many organic compounds are volatile liquids)
- to avoid fires (many organic compounds are flammable)

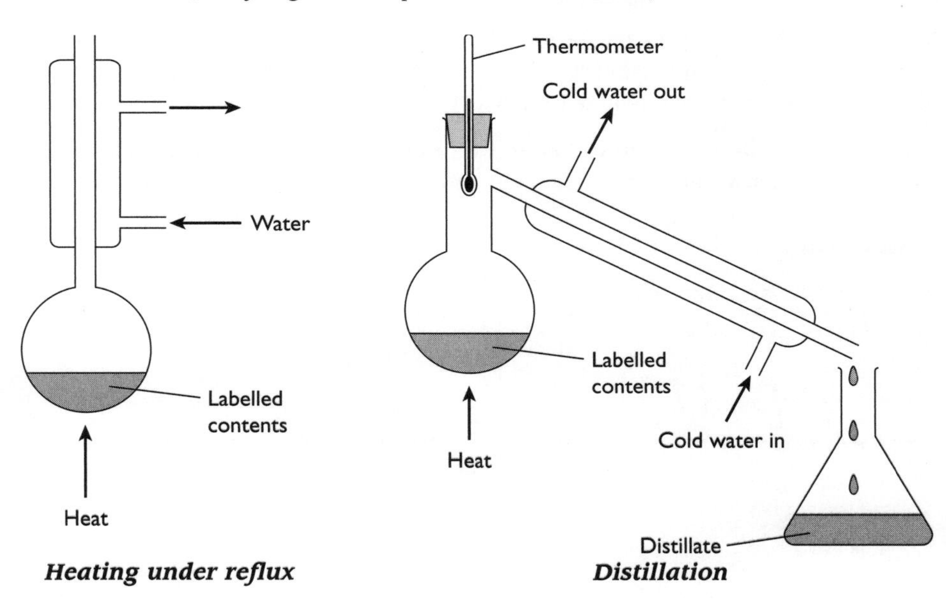

Heating under reflux *Distillation*

Distillation

You met this in **Elements from the sea**.

This is used as the final step in the purification of an organic liquid. The (slightly impure) liquid is distilled and the distillate collected at the boiling point required.

When drawing either apparatus, note the following:

- Label the heat source and the mixture you are heating.
- Make it obvious that there is a clear path through the apparatus and *do not stopper the top of a reflux apparatus.*

- Show the water connections correctly. Water goes in at the lower connection so that it fills the apparatus, rather than rushing through it.
- Do not (carelessly) leave any gaps through which vapour could escape. (Thus, do not try to represent Quickfit apparatus with ground-glass joints.)

Condensation polymers

You meet this topic in **The materials revolution**.

Condensation reactions

A **condensation** reaction is defined as one where two molecules join together to make a larger molecule, with the elimination of a small molecule (commonly, but not always, water). **Polymerisation** occurs when many small molecules (**monomers**) join to form a larger molecule — the **polymer**. Thus for condensation polymerisation to occur the monomers must have two reactive groups that can form a polymer chain by condensation reactions.

Note that, in addition polymerisation, the monomer has the same empirical (ratio) formula as the polymer, whereas this is not the case for condensation polymers because a small molecule (e.g. water) is lost.

The two reactions you need to know are **esterification** and **amide formation**. Examples are shown in the table on page 16.

Polyesters

A *di*carboxylic acid and a *di*ol can react to form a *poly*ester. The reaction continues at either end of the chain.

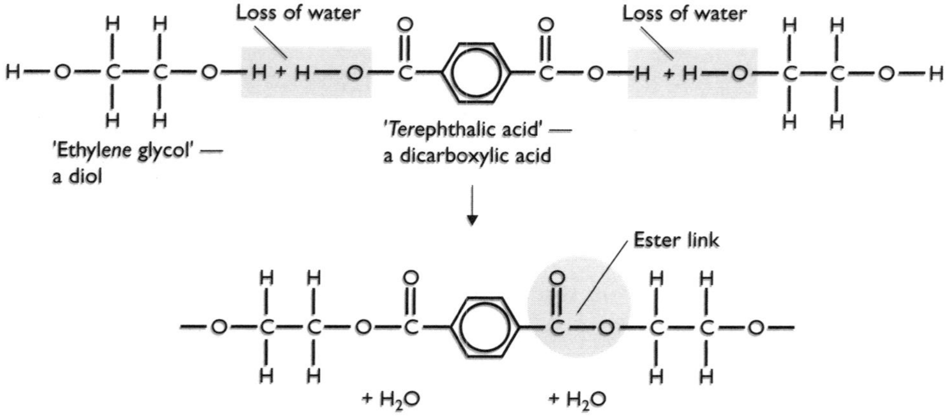

Loss of water

Loss of water

'Ethylene glycol' — a diol

'*Terephthalic acid*' — a dicarboxylic acid

Ester link

+ H_2O + H_2O

Polyester ('*terylene*')

Tip You do not need to know the old names given in quotation marks — they just illustrate how the name 'terylene' came about.

Terylene was designed on purpose, not discovered by accident. The repeating unit is shown below:

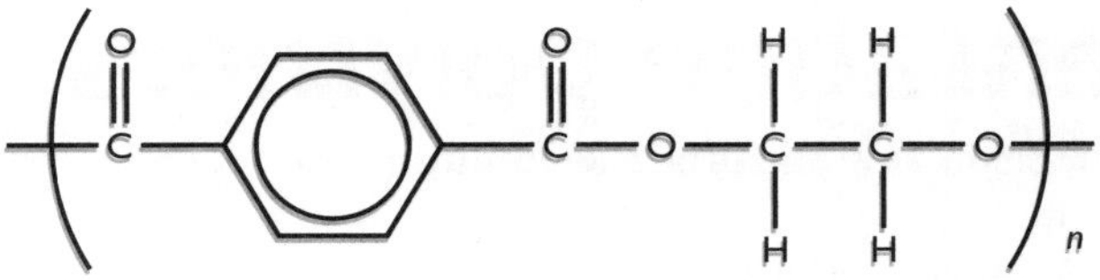

$$\left[-CO-C_6H_4-CO-O-CH_2-CH_2-O- \right]_n$$

Tip Note the 'half' ester group at each end.

Names

Diols are named thus: pentane-1,3-diol ($CH_3CH_2CH(OH)CH_2CH_2OH$)

Dicarboxylic acids are named by the total number of carbon atoms, for example butanedioic acid ($HOOCCH_2CH_2COOH$). Note that the 'e' from the alkane reappears before the consonant 'd' (butanoic acid but butanedioic acid).

Polyamides

In the following reaction a diamine and a dicarboxylic acid form a polyamide, called **nylon**. The reaction can continue at either end of the chain.

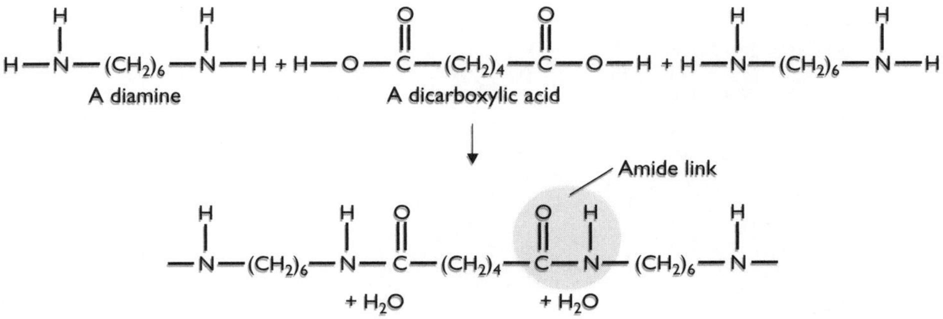

$$H-NH-(CH_2)_6-NH-H + H-O-CO-(CH_2)_4-CO-O-H + H-NH-(CH_2)_6-NH-H$$

A diamine — A dicarboxylic acid

$$\downarrow$$

$$-NH-(CH_2)_6-NH-CO-(CH_2)_4-CO-NH-(CH_2)_6-NH-$$

$+ H_2O$ $+ H_2O$

Amide link

The repeat unit is:

$$\left[-NH-(CH_2)_6-NH-CO-(CH_2)_4-CO- \right]_n$$

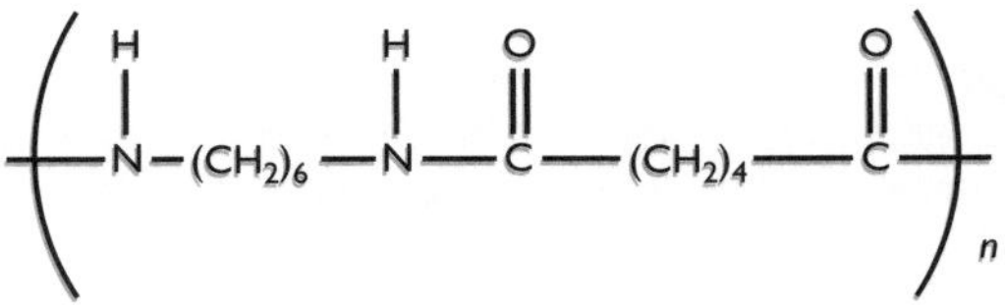

The amino acid shown below has both the amine group and the acid group, so it would polymerise with itself (under the right conditions) to form a similar nylon.

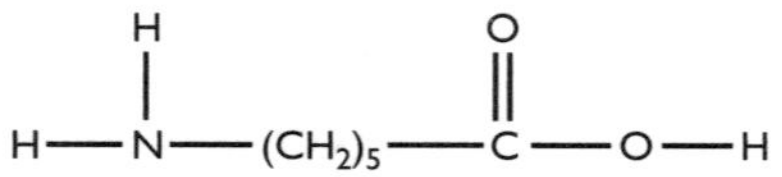

$$\mathrm{H{-}N(H){-}(CH_2)_5{-}C({=}O){-}O{-}H}$$

Names

Diamines are named using the prefix 'diamino' — for example, $NH_2CH_2CH_2NH_2$ is 1,2-diaminoethane.

Hydrolysis

This is the reverse of condensation.

Esters

If an ester is heated under reflux with sodium hydroxide, the sodium salt of the acid and an alcohol are formed. Note that if moderately concentrated HCl is used for hydrolysis, the acid and the alcohol are produced:

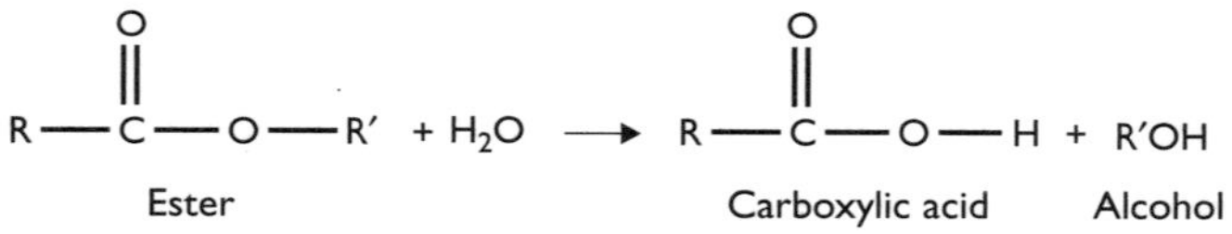

$$\underset{\text{Ester}}{\mathrm{R{-}C({=}O){-}O{-}R'}} + \mathrm{H_2O} \longrightarrow \underset{\text{Carboxylic acid}}{\mathrm{R{-}C({=}O){-}O{-}H}} + \underset{\text{Alcohol}}{\mathrm{R'OH}}$$

So a *polyester* will hydrolyse to the *dicarboxylic acid* and the *diol*.

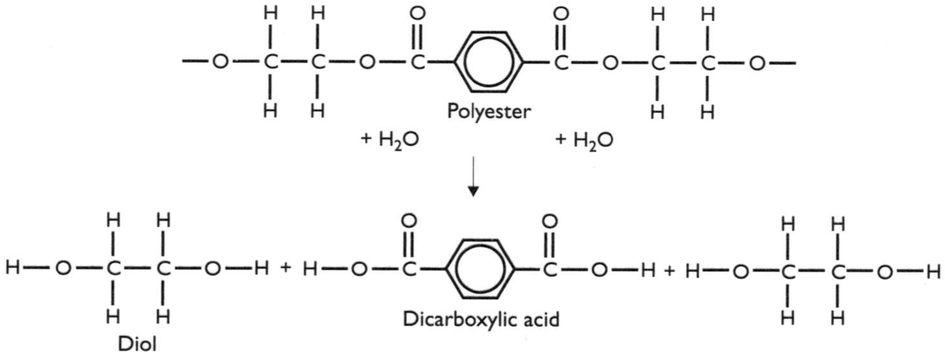

$$\underset{\text{Polyester}}{\mathrm{{-}O{-}CH_2{-}CH_2{-}O{-}C({=}O){-}C_6H_4{-}C({=}O){-}O{-}CH_2{-}CH_2{-}O{-}}} \quad + \mathrm{H_2O} \quad + \mathrm{H_2O}$$

$$\downarrow$$

$$\underset{\text{Diol}}{\mathrm{H{-}O{-}CH_2{-}CH_2{-}O{-}H}} + \underset{\text{Dicarboxylic acid}}{\mathrm{H{-}O{-}C({=}O){-}C_6H_4{-}C({=}O){-}O{-}H}} + \mathrm{H{-}O{-}CH_2{-}CH_2{-}O{-}H}$$

Amides

If an amide is heated under reflux with moderately concentrated hydrochloric acid, the *salt of the amine* and the *carboxylic acid* are formed (heating with moderately concentrated alkali produces the *amine* and the *salt of the acid*).

$$R-\overset{\overset{\large O}{\|}}{C}-O-\overset{\overset{\large H}{|}}{N}-R' + H_2O \xrightarrow{\text{Acid hydrolysis}} R-\overset{\overset{\large O}{\|}}{C}-O-H + H-\overset{\overset{\large H}{|}}{\underset{\underset{\large H}{|}}{N^+}}-R'$$

Amide — Carboxylic acid — Amine salt

So, in acid conditions, *polyamides* hydrolyse to *dicarboxylic acids* and *diamine salts*.

$$-\overset{\overset{\large H}{|}}{N}-(CH_2)_6-\overset{\overset{\large H}{|}}{N}-\overset{\overset{\large O}{\|}}{C}-(CH_2)_4-\overset{\overset{\large O}{\|}}{C}-\overset{\overset{\large H}{|}}{N}-(CH_2)_6-\overset{\overset{\large H}{|}}{N}-$$

$+ H_2O$ $\quad$ $+ H_2O$

Nylon

↓ Acid hydrolysis (e.g. HCl)

$$H-\overset{\overset{\large H}{|}}{\underset{\underset{\large H}{|}}{N^+}}-(CH_2)_6-\overset{\overset{\large H}{|}}{\underset{\underset{\large H}{|}}{N^+}}-H + H-O-\overset{\overset{\large O}{\|}}{C}-(CH_2)_4-\overset{\overset{\large O}{\|}}{C}-O-H + H-\overset{\overset{\large H}{|}}{\underset{\underset{\large H}{|}}{N^+}}-(CH_2)_6-\overset{\overset{\large H}{|}}{\underset{\underset{\large H}{|}}{N^+}}-H$$

Diamine salt (e.g. chloride) — Dicarboxylic acid

Properties of polymers

Intermolecular bonds

The three types of intermolecular bonds are, in order of strength:
hydrogen bonds > permanent dipole–permanent dipole bonds > instantaneous dipole–induced dipole bonds

- **Hydrogen bonds** occur when molecules or polymers have –OH or –NH groups — for example, nylon.
- **Permanent dipole–permanent dipole bonds** are the strongest intermolecular force in molecules or polymers that have O or N (*without* O–H or N–H) or halogen atoms — for example, polyesters and PVC.
- **Instantaneous dipole–induced dipole bonds** are present in *all* molecules, but they are usually the strongest force only when the others are absent. These are present in poly(ethene) and poly(propene).

The strength of nylon depends partly on the hydrogen bonds between adjacent chains. The polymer **Kevlar** has been designed to maximise hydrogen bonds between the chains. It is so strong that it is used for tyres and bullet-proof vests. Its structure is:

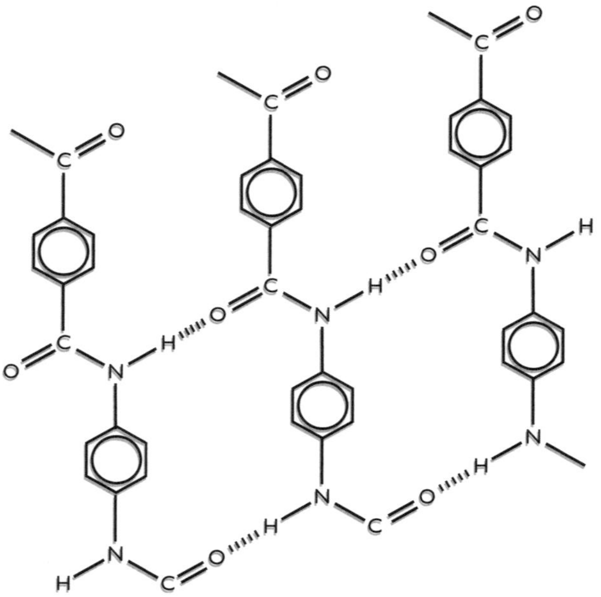

The Kevlar chains are held flat by the lack of rotation of the benzene rings. The hydrogen bonds hold the flat molecules into sheets, which accounts for Kevlar's strength.

Effect of heat on polymer structures

In **The polymer revolution** you learnt about **thermoplastics** and **thermosets**, the latter having covalent bonds between the chains so that they do not soften or melt on heating. Thermoplastics have weaker intermolecular bonds between their chains and some more subtle changes occur when they are heated.

At low enough temperatures, all thermoplastics are *glassy*. The chains cannot move over one another and, if enough force is applied, the polymer material simply breaks.

As polymers are heated, they reach their glass transition temperature, T_g, above which the material becomes flexible. This temperature varies with the structure of the polymer. Eventually, on further heating, the polymers soften and 'melt'. The temperature at which they do this is called T_m.

Polymers that have to be rigid are designed so that their T_g is above room temperature — for example, unplasticised PVC used for window frames. Other polymers have to be more flexible and are designed so that room temperature is between T_g and T_m.

T_g is higher when the **intermolecular** bonds are stronger (or have more effect because the chains are closer), holding the chains more tightly together.

Varying the properties

Chemists are called upon to produce polymers for different and specific functions. First, a polymer that has properties close to those required is selected. The properties of a polymer can then be changed by:

- **cold drawing:** when a polymer is gently pulled, 'necking' occurs and the polymer chains line up to produce more crystalline regions.

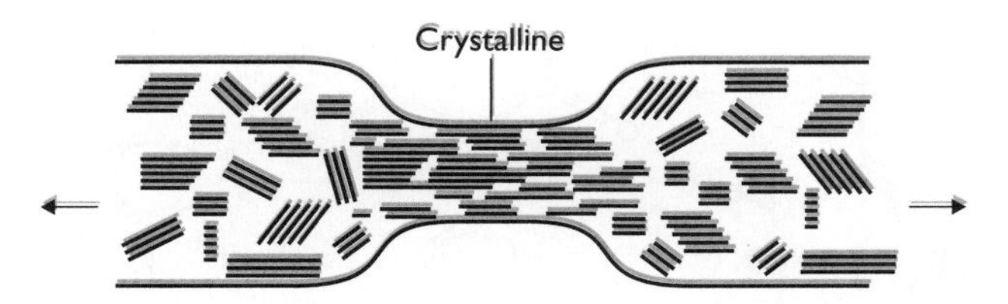

In a crystalline area of polymer, the chains lie closer together and thus the intermolecular bonds have more effect.

- **increasing chain length:** this increases the total number of intermolecular bonds (whatever they are) and leads to more bonding between chains, hence greater strength.
- **copolymerising:** this incorporates monomers with different functional groups into the structure and this changes the polymer structure and properties. An example is the copolymerising of a little ethenyl ethanoate with chloroethene (the monomer of PVC).

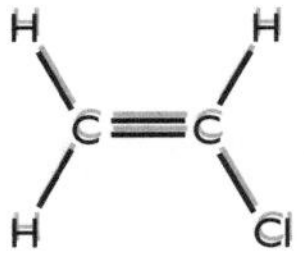

Chloroethene

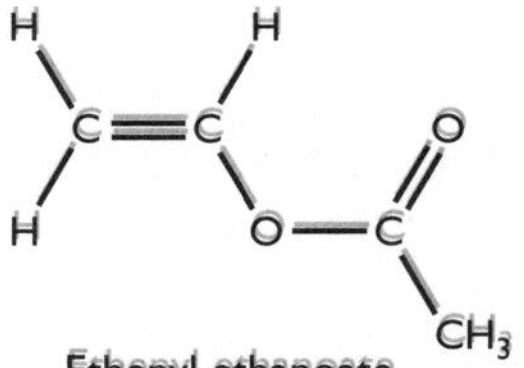

Ethenyl ethanoate

Tip You do not need to learn the formula of ethenyl ethanoate.

Both monomers contribute permanent dipole–permanent dipole bonds to the polymer, but ethenyl ethanoate has larger side groups. The occasional large side-groups cause the chains to pack together less well, so the intermolecular bonds are weaker and the chains can move past each other more easily. The polymer material is more flexible and has a lower T_g and T_m. The copolymer is called 'plasticised' PVC.

- **use of plasticisers:** see above. It is also possible to plasticise PVC (and other polymers) by adding a **plasticiser** molecule that simply gets between the chains and has the same effect by pushing them apart.

How to answer questions on polymers

Example: Which of two polymers A and B is more flexible (or has lower tensile strength)?	Begin by saying that polymer A is flexible (has lower tensile strength) because its chains can move past each other more easily.
Example: Which of the two polymers A and B has a lower T_g (or T_m)?	Begin by saying that polymer A has a lower T_g (or T_m) because its chains can move past each other more easily. Less energy is required to make the chains move.

Then explain why the chains can move past each other more easily. Remember to compare properties each time. Three alternative approaches are given below:

- Polymer A has instantaneous dipole-induced dipole bonds whereas polymer B has hydrogen bonds because... (explanation of intermolecular bonds in each polymer). Hydrogen bonds are stronger than instantaneous dipole-induced dipole bonds or B has more hydrogen bonds than A because...
- Polymer A has shorter chains so the intermolecular bonds (name them) have less effect.
- Polymer A is less crystalline than polymer B. Thus the chains do not lie as closely together and the intermolecular bonds (name them) between chains have less effect.

Look for clues in the question as to which vital property to compare.

- 'In terms of intermolecular bonds' means intermolecular bonds should be compared.
- 'The chain lengths are similar' means chain lengths should not be compared and intermolecular bonds probably should be compared.

Proteins

Proteins are covered in **The thread of life**. They are naturally occurring polymers formed by condensation polymerisation of amino acids.

Amino acids

These are the building blocks of proteins. There are 20 amino acids that make up our proteins and they all have the same type of structure:

$$H_2N-\overset{\displaystyle R}{\underset{\displaystyle H}{\overset{|}{\underset{|}{C}}}}-COOH$$

In this context, the R group is *not* necessarily an alkyl group. For example:

- If R is –H, the amino acid is called **glycine**.
- If R is $-CH_2OH$, the amino acid is called **serine**.

Tip You do not need to learn these names.

Properties of amino acids

Acid–base

The $-NH_2$ group is basic and the –COOH group is acidic:

$-NH_2 + H^+ \rightarrow NH_3^+$ (this occurs when acids are added to amino acids)

$-COOH \rightarrow -COO^- + H^+$ ($-COOH + OH^- \rightarrow -COO^-$ occurs when alkalis are added)

Both these reactions occur within an amino acid molecule, forming a **zwitterion**.

$$^+H_3N-\overset{\displaystyle R}{\underset{\displaystyle H}{\overset{|}{\underset{|}{C}}}}-COO^-$$

This explains why solid amino acids have high melting points and why they are very soluble in water. The solution is nearly neutral unless there are extra –COOH or $-NH_2$ groups.

Optical isomerism

All the amino acids (except glycine) have *four different groups* around the central carbon and thus show **optical isomerism** (see p. 21).

Protein structure

Primary structure

The primary structure consists of amino acids joined together by secondary amide bonds (see p. 20). In the context of proteins, these bonds are called **peptide** links. The diagram below shows glycine and alanine (R group, CH_3) joining together as part of a protein structure. This is a **condensation reaction** — water is eliminated.

Glycine Alanine

Peptide link $+ H_2O$

Part of the primary structure of a protein

Two amino acids combined together are called a **dipeptide** (even though there is only *one* peptide link).

The *sequence* of amino acids in the protein structure is called its **primary structure**. The primary structure is characteristic of the particular protein and determines the way the protein folds — for example, whether it is designed to be a fibrous *structural* protein (like hair or fingernails) or a globular protein such as an **enzyme** (biological catalyst).

The amino acids present in a protein can be identified as follows:

- Heat a sample of the protein under reflux with moderately concentrated hydrochloric acid. This **hydrolyses** it into its individual amino acids.
- Spot the resulting mixture onto chromatography paper.
- Run the chromatogram using a suitable *solvent*.
- Use a **locating agent** (ninhydrin), followed by warming in an oven, to make the spots visible.

The amino acids can be identified by running known amino acids alongside and comparing which move to the same height, or by measuring R_f values.

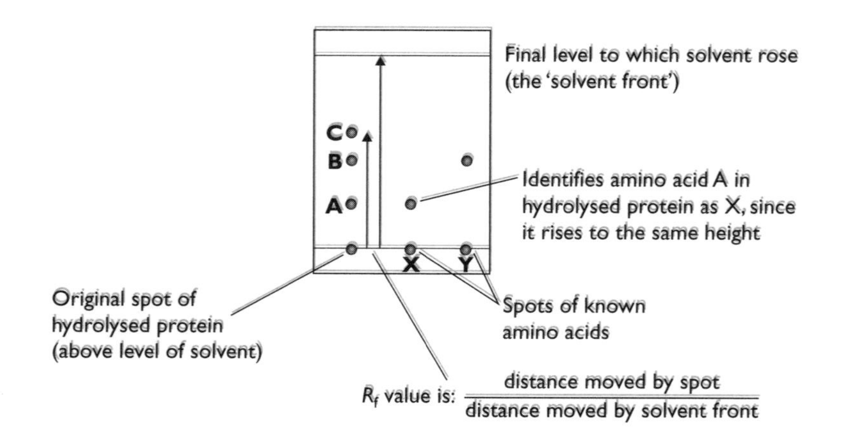

Secondary structure

This describes how the amino acid chain folds. There are two possible folding patterns, the **helix** and the **sheet**. Secondary structures are held together by *hydrogen bonds* between the –C=O groups and –N–H groups as shown in the diagrams below.

Helix

Sheet

Tertiary structure

This is how helical portions and/or sheets fold up further to give the overall shape of the protein. Fibrous proteins, such as the hair protein keratin, have a tertiary structure that consists of many helices wound together, in a similar way to pieces of thread being wound together to make string. Globular proteins such as enzymes have a mixture of helices and sheets in their structure:

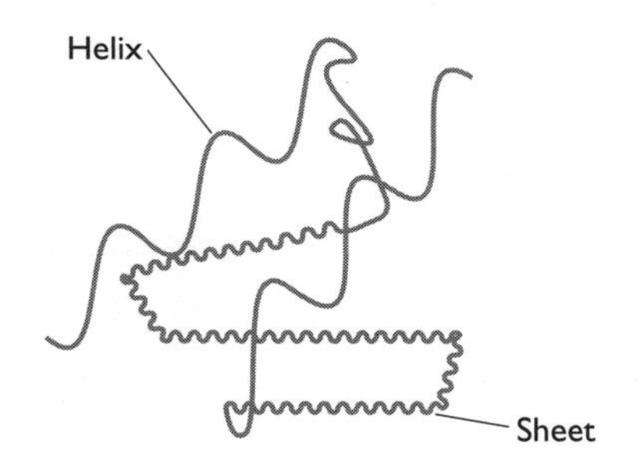

The tertiary structure is held together by various types of interaction, which are shown in the diagram below. However, this is schematic — you would not expect to find so many different bonds so close together.

S S COO^- NH_3^+ O H O H CH_3 CH_3

Covalent bonds 'disulfide bridges' | Ionic interactions | Hydrogen bonds | Instantaneous dipole–induced dipole bonds

Enzymes

These are **biological catalysts** present in all living systems. They are used industrially in a variety of important processes — for example, brewing, wine-making and cheese-making. They are also used in washing powders.

The tertiary structure of an enzyme is folded to form a cleft in the molecule with a specific shape into which the **substrate** molecule fits. (The substrate is the molecule whose reactions are being catalysed.) This cleft is called the **active site**. Within the active site, important R groups on the protein chain are held in position by the tertiary structure and bind the substrate to the enzyme.

The substrate binds by weak intermolecular bonds to the active site forming an **enzyme–substrate complex**. Here bond enthalpies are changed and reactions occur that normally require much greater activation enthalpies.

The products no longer fit the active site and bind less well, so they leave. The enzyme is then available to catalyse the reaction of a further substrate molecule.

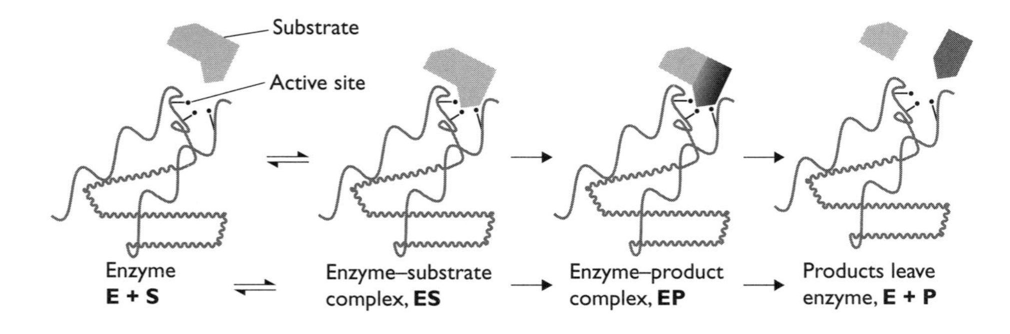

Characteristics of enzyme reactions

Enzymes are:

- *specific*. In order to fit and bind onto the active site, the substrate molecule must be exactly the right shape. Where the substrate is a chiral molecule, often only the reaction of one **enantiomer** is catalysed.
- *sensitive to pH*. pH affects the ratio of $-COO^-$ to NH^+ present in R groups. Small changes of pH can affect the charges on the R groups present in the active site itself, resulting in a decrease in activity. A change of pH of more than two units causes the tertiary structure of the enzyme to break down, with the loss of the active site. This is called **denaturation**.

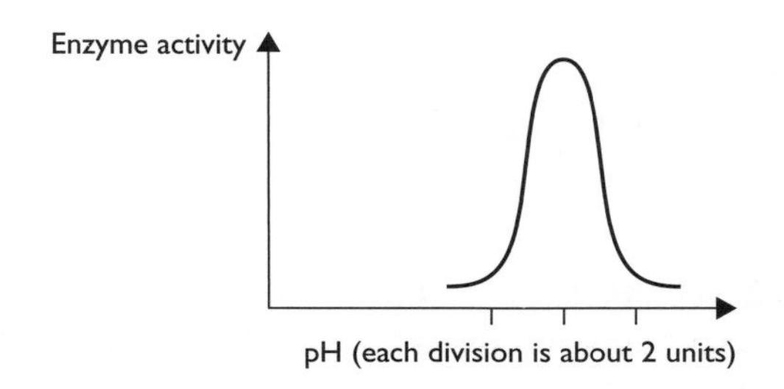

- *sensitive to temperature*. As the temperature of an enzyme-catalysed reaction is raised, the reaction gets faster, as with all other reactions. However, above a certain temperature, the activity rapidly falls to zero as the enzyme becomes denatured. In this case, the molecular agitation caused by the increasing temperature breaks the hydrogen bonds holding both the secondary and tertiary structures together. The active site is broken up and the enzyme no longer functions as a catalyst.

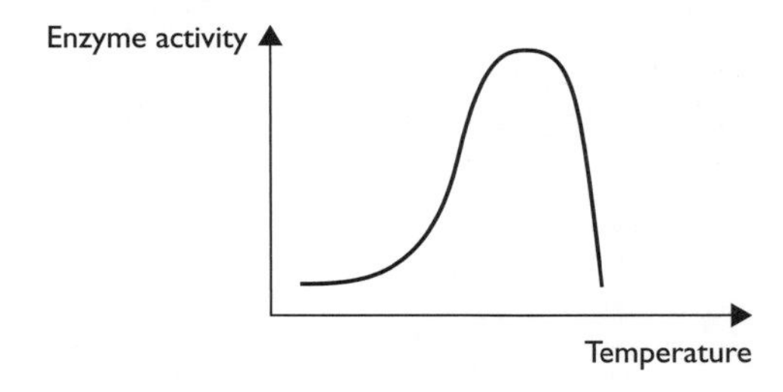

- *inhibited* by substances that bind to the active site but do not react. Substances that resemble substrate molecules often inhibit enzymes. Substrate and inhibitor compete for occupancy of the active site. However, once in the active site, inhibitors do not react. Substrate molecules cannot enter, so the reaction rate slows down. For example, the enzyme succinate dehydrogenase catalyses the oxidation of butanedioic (succinic) acid by removing hydrogen:

$HOOC–CH_2–CH_2–COOH \rightarrow HOOC–CH=CH–COOH$

The enzyme is inhibited by propanedioic acid ($HOOC–CH_2–COOH$), which can bind to the active site but has no $CH_2–CH_2$ link from which hydrogen can be removed.

At high substrate concentrations, enzyme catalysed reactions are **zero-order** with respect to substrate concentration (see p. 43).

DNA

You meet DNA in in **The thread of life**.

Structure

DNA stands for deoxyribonucleic acid. This molecule carries the code to make proteins. Since enzymes are proteins, this means that DNA controls all our body functions.

DNA has a double helix structure. (Note that it is not a protein.) Each strand consists of a **sugar–phosphate** backbone (the sugar being deoxyribose). Attached to each sugar is a **base** denoted by the letters T, A, G and C. These hydrogen-bond in pairs (T=A and C≡G), which is how the two strands of the helix are held together.

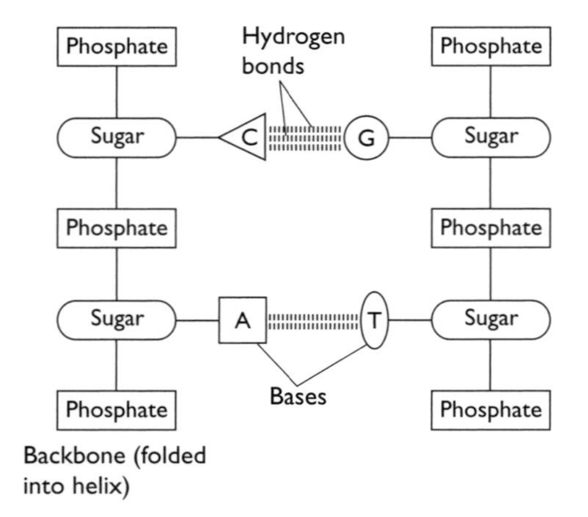

You need to know this outline structure, but you also need to be able to use the monomer structures on the *Data Sheets* to give the chemistry. The monomers join together by condensation reactions:

$$-OH + HO- \rightarrow -O- + H_2O \quad \text{or} \quad -NH + HO- \rightarrow -N- + H_2O$$

Bases combine by condensing the H on an N-H with the right-hand -OH on the sugar, as drawn below. The -OH groups on the other side of the sugar each condense with a phosphate group, continuing the chain:

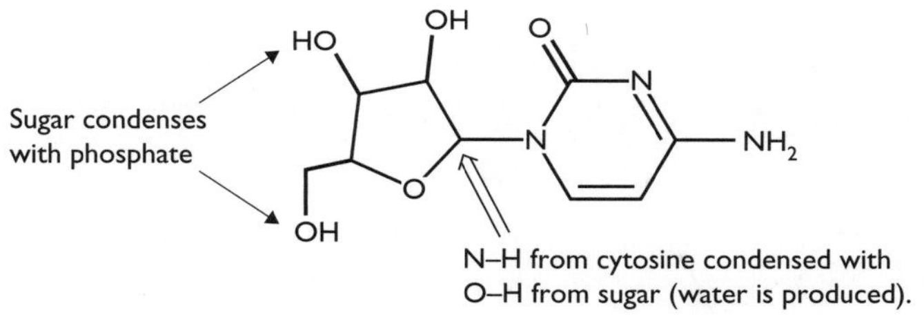

Another example is adenine condensed with deoxyribose, itself condensed with phosphate:

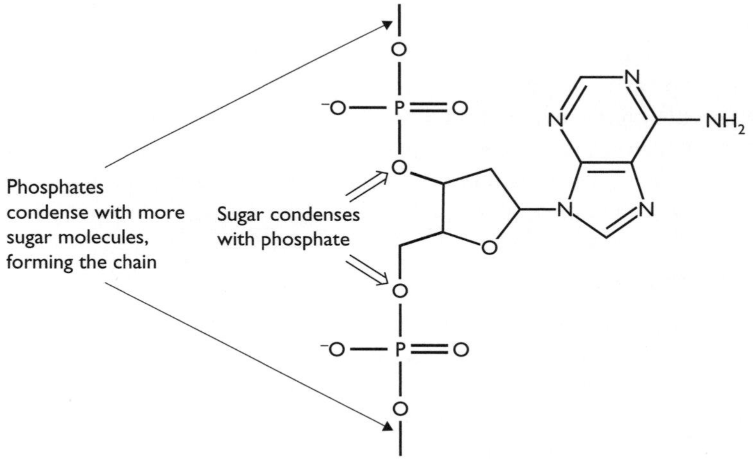

The fact that the bases pair up is important in cell division. When a cell divides, the DNA double helix untwists and each strand goes to one of the new cells. Strands of DNA can form the template for the other strand. This is called **replication** and this is the way the genetic code is passed on as cells divide.

You need to be able to draw the bases paired together. Although you have the structures from the *Data Sheets*, you need to practise to see how these are manipulated and fit together. One way of doing this is shown in the diagram below.

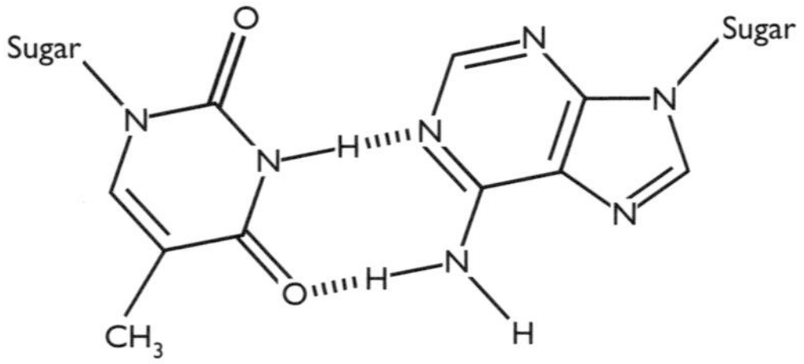

Thymine (rotated clockwise) and adenine (rotated anticlockwise, then 'flipped' vertically)

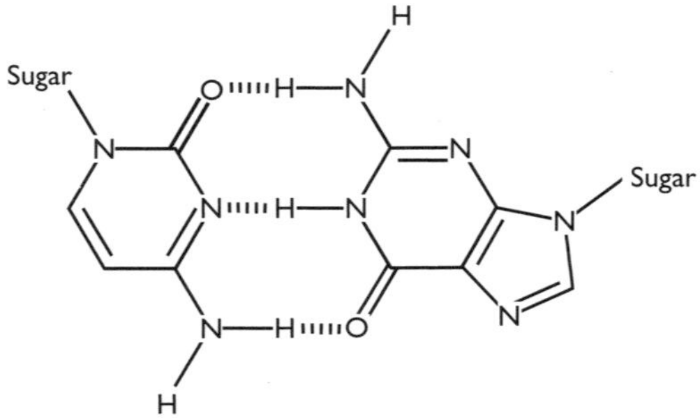

Cytosine (rotated clockwise) and guanine (rotated anticlockwise, then 'flipped' vertically)

Tip Learn the following and practise drawing out structures that show:

- the fundamental 'base–sugar–phosphate' block diagram above
- which atoms on each monomer structures are involved in condensation reactions
- how the hydrogen bonds form between each base pair

Protein synthesis

RNA stands for **ribonucleic acid**. RNA has the base U (uracil) instead of T; U pairs with A. It also has ribose as the sugar instead of deoxyribose. During protein synthesis, the DNA in the nucleus makes a strand of **messenger RNA** in a similar way to that in which the DNA replicates itself. The messenger RNA carries the code from the DNA in the nucleus to the *ribosomes* which are in the cytoplasm outside the nucleus. Protein synthesis occurs in the ribosomes.

Messenger RNA is a single-strand molecule. Each group of *three bases* is called a **codon** because it codes for an amino acid. In the cytoplasm of the cell, each amino acid has its own **transfer RNA** molecule with an **anticodon** that has bases complementary to those of a particular codon. The transfer RNA molecules bring the amino acids to the ribosomes. As the ribosome moves along the messenger RNA, each anticodon forms hydrogen bonds with the corresponding codon. In this way the primary structure of the protein is assembled in the right order, so that the protein can fold into its correct shape.

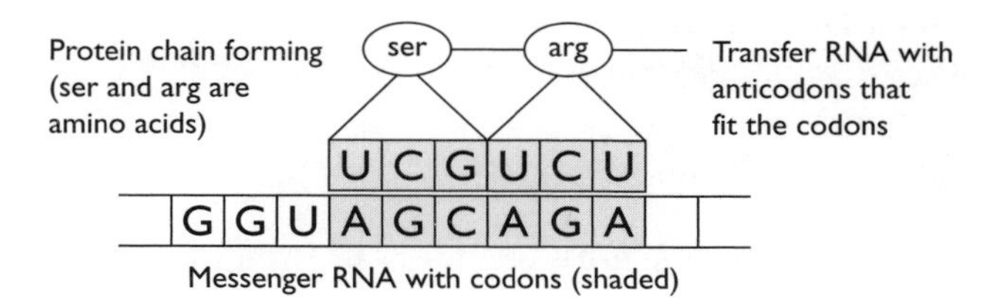

Tip You need to be able to describe and explain this mechanism, but there is a diagram in the *Data Sheets* to help you. Be sure to study this.

DNA databases

Since no two people (apart from identical twins) have exactly the same DNA, DNA sequences can be used to identify individuals. A trace of blood, semen or skin is all that is needed to extract a sample of DNA. The DNA is chemically divided into fragments. Much of our DNA is 'junk' and does not take part in synthesising proteins. Areas of junk DNA differ most between individuals and can best be used for identification.

Many countries keep in a database the DNA profiles of those convicted of crime. Some (e.g. England) also keep the DNA samples of those arrested, even if they are not charged or if they are acquitted.

Arguments for DNA databases

- Databases are useful in solving crimes now and in the future.
- Innocent people have nothing to fear.

Arguments against DNA databases

- It is an offence against the human rights of the individual.
- Governments might one day change and become less benign.

Tip You should learn these arguments (or others that you have discussed).

DNA testing

Is it a good idea to test the DNA of individuals, including babies and even unborn children? The DNA can be analysed to show the likelihood of genetic disease or the possibility of developing certain illnesses later in life.

Arguments for DNA testing

- It would enable early warning of a possible condition and allow action to be taken.
- It is an individual's right to know this information.

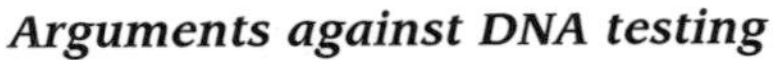

Arguments against DNA testing

- If the condition were untreatable, it might cause years of unnecessary worry.
- If people knew they were likely to have a certain disease, they would have to declare it to insurance companies etc.

Tip You should learn these arguments (or others that you have discussed).

Reaction rates

You meet reaction rates (in connection with enzymes) in **The thread of life**. **Rate of reaction** at a given moment is defined as the change in concentration divided by the time taken for the change.

Effect of concentration

Rate equation and rate constant

The rate equation for a reaction A + B → products is:

$$\text{Rate} = k[\text{A}]^a\,[\text{B}]^b$$

where k is the **rate constant**, square brackets represent concentrations in mol dm^{-3} and a and b are the **orders of reaction** with respect to A and B respectively. (The overall order of reaction is $a + b$.)

Units of the rate constant

The units of rate are mol dm^{-3} s^{-1}. Thus for an overall first-order reaction:

k = Rate/[reagent] Units = mol dm^{-3} s^{-1}/mol dm^{-3} = $\mathbf{s^{-1}}$

For a second-order reaction:

k = Rate/$[\text{reagent}]^2$ Units = mol dm^{-3} s^{-1}/mol^2 dm^{-6} = $\mathbf{mol^{-1}\ dm^3\ s^{-1}}$

Tip Learn these units if you must, but it is better to learn how to work them out.

Measuring orders of reaction

Orders of reaction have nothing to do with the overall balanced chemical equation for the reaction because this usually represents a series of steps. Orders must be measured by experiment.

Method 1

This involves measuring the time for a small amount of the reaction to occur. The reciprocal of the time taken (1/time) is proportional to the *initial* rate, which applies when the reactants' concentrations have their starting values.

For example, if three reagents A, B and C react and the initial rates are measured with different starting concentrations:

Tube	A	B	C	Relative rate
1	1	1	1	1
2	2	1	1	1
3	1	2	1	2
4	1	1	2	4

- Comparing tube 2 with tube 1 — the concentration of A has doubled and the rate has remained the same. This means that the order of reaction is *zero* with respect to A.
- Tubes 1 and 3 show that the rate doubles when the concentration of B doubles, so the order is *first* with respect to B.
- Tubes 1 and 4 show that the order is second with respect to C as the rate goes up by four when the concentration of C doubles.

Method 2

This involves following the reaction over the course of its progress. The reaction is followed by either:

- measuring the concentration of a reactant or product against time or
- by measuring a property that is proportional to the concentration

One reagent is *limiting* (in smaller concentration than the others) and a graph is plotted of this reactant's concentration against time. If the *half-life* (see below) is constant, the reaction is first-order with respect to the limiting reagent (if the graph is a straight line, it is zero-order; if successive half-lives increase it is second-order).

For example, the graph shows the variation of concentration of X with time. This reaction is first-order with respect to X.

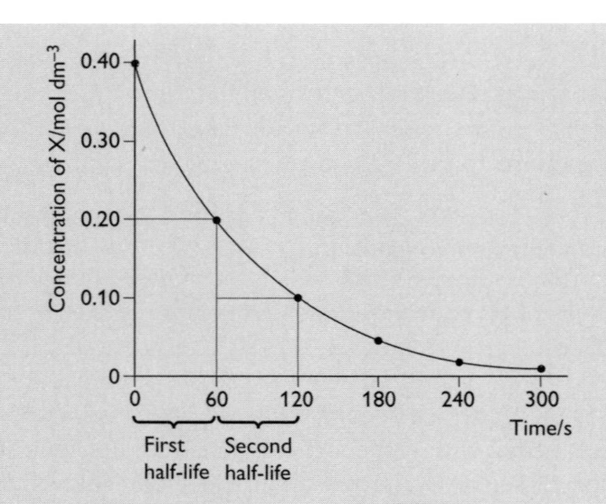

When plotting graphs and working out half-lives:

- choose scales carefully to make plotting easy
- label axes fully (don't forget powers of ten and units)
- plot points carefully and accurately, in pencil
- carefully draw the best smooth curve through the points in pencil
- show half-life working on the graph as above

Following reactions

In both of the above methods, it is necessary to follow some change that tells us how far the reaction has gone. Ways of following reactions include:

- sampling the reaction mixture and titrating to find the concentration of a reagent
- following the pH, using a pH meter, if an acid is produced or used up
- following the reaction with a colorimeter if a coloured compound is made or used up
- measuring the volume change if a gas is produced
- measuring mass changes if a gas is produced

Reaction mechanisms

Many reactions occur in several steps. The slowest step (like a 'bottleneck') is called the **rate-determining step**. If a reactant is used in fast step *after* the rate-determining step, its concentration will not affect the overall rate of the reaction, i.e. the overall reaction is **zero-order** with respect to that reactant.

Enzyme kinetics

Enzyme reactions occur as follows:

Step 1: E + S ⇌ ES

Step 2: ES → E + P

(see p. 34)

At *low* substrate concentrations, *step 1* is rate-determining. Thus the reaction is **first order** with respect to both enzyme and substrate. However, enzymes are molecules with large M_r so their molar concentrations can be very small. As the substrate concentration rises, it becomes very much greater than the enzyme concentration. The enzyme is then said to be 'saturated' with all active sites occupied, i.e. it spends almost all its time as ES. Under these conditions, *step 2* becomes rate-determining and the reaction is **zero order** with respect to the substrate concentration, though still first order with respect to enzyme concentration. The graph below illustrates this.

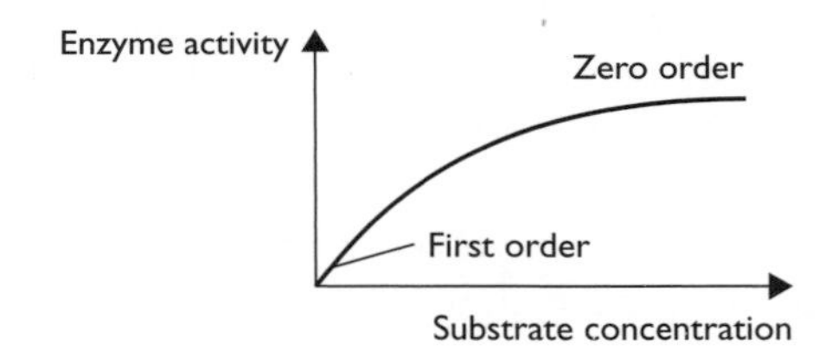

Transition metals

Electronic structure

Transition elements lie in the ***d*-block** of the periodic table, where the ***d*-subshell** is the one being filled. Given the atomic number, you are expected to be able to work out the electron structures of the *first row* of *d*-block elements, from scandium to zinc. (You can look up the atomic numbers in the periodic table in the *Data Sheets*.)

After argon ($1s^2\ 2s^2\ 2p^6\ 3s^2\ 3p^6$) the next element (potassium) starts to fill the $4s$-subshell, since the $3d$-subshell is slightly above it in terms of energy. However, after calcium ($1s^2\ 2s^2\ 2p^6\ 3s^2\ 3p^6\ 4s^2$), the $3d$-subshell starts to fill. Thus, the electron structure of scandium can be written as either $1s^2\ 2s^2\ 2p^6\ 3s^2\ 3p^6\ 3d^1\ 4s^2$ or [Ar] $3d^1\ 4s^2$, where [Ar] represents the electronic structure of argon. Alternatively, the 'electron box' representation can be used. Each box represents an **orbital** (named from the time when people thought that electrons orbited round the nucleus):

	Atomic number	*spdf* description	Electron box description 3*d*-orbitals	4*s*-orbital	Notes
Sc	21	[Ar] $3d^14s^2$	↑ | | | |	↑↓	Sc^{3+} is the only ion with [Ar] structure*
Ti	22	[Ar] $3d^24s^2$	↑ | ↑ | | |	↑↓	
V	23	[Ar] $3d^34s^2$	↑ | ↑ | ↑ | |	↑↓	
Cr	24	[Ar] $3d^54s^1$	↑ | ↑ | ↑ | ↑ | ↑	↑	Note: does not follow pattern**
Mn	25	[Ar] $3d^54s^2$	↑ | ↑ | ↑ | ↑ | ↑	↑↓	
Fe	26	[Ar] $3d^64s^2$	↑↓ | ↑ | ↑ | ↑ | ↑	↑↓	
Co	27	[Ar] $3d^74s^2$	↑↓ | ↑↓ | ↑ | ↑ | ↑	↑↓	
Ni	28	[Ar] $3d^84s^2$	↑↓ | ↑↓ | ↑↓ | ↑ | ↑	↑↓	
Cu	29	[Ar] $3d^{10}4s^1$	↑↓ | ↑↓ | ↑↓ | ↑↓ | ↑↓	↑	Note: does not follow pattern**
Zn	30	[Ar] $3d^{10}4s^2$	↑↓ | ↑↓ | ↑↓ | ↑↓ | ↑↓	↑↓	Zn^{2+} is the only ion with [Ar] $3d^{10}$ structure*

** The definition of a transition metal is an element that forms at least one ion with a partially filled d-subshell of electrons. Scandium and zinc do not meet this definition. Therefore, they are d-block elements but not transition metals.*

*** d^5 and d^{10} are particularly stable electron arrangements. Thus, chromium and copper have the electron arrangements as shown above with $4s^1$, not $4s^2$.*

Ions are formed by losing the 4*s*-electrons first, then the *d*-electrons. For example, Fe^{2+} is [Ar] $3d^6$ and Fe^{3+} is [Ar] $3d^5$. Cu^{2+} is [Ar] $3d^9$.

Typical properties

Variable oxidation state

The only ion formed by magnesium is Mg^{2+}, because the most stable electron arrangement is [Ne]. However, within the 3*d*- and 4*s*-subshells there are usually several stable electron arrangements, resulting in several oxidation states. The only oxidation states you need to learn are those of iron (+2 and +3) and copper (+1 and +2).

Naming compounds

Systematic names for metal compounds use the *Stock* nomenclature. The oxidation state of the cation is given after the metal name, if more than one oxidation state exists. The oxidation state of the metal or non-metal in a complex anion is also given. For example, CuS is copper(II) sulfide; $Fe_2(SO_4)_3$ is iron(III) sulfate(V) and $KMnO_4$ is potassium manganate(VII).

Tip You do not need to be able to name complex ions such as $[Cu(H_2O)_6]^{2+}$ or $[CuCl_4]^{2-}$.

Formation of complexes

A **complex** is a metal atom or ion surrounded by **ligands**. Ligands are negative ions or neutral molecules with a lone pair of electrons that they can donate to the metal ion to form complexes. Some examples are given in the table below.

Ligand	Example	Colour	Charge on metal ion	Charge on complex
H_2O	$[Fe(H_2O)_6]^{3+}$	Orange	+3	+3 (Ligand is neutral)
H_2O	$[Fe(H_2O)_6]^{2+}$	Pale green	+2	+2 (Ligand is neutral)
H_2O	$[Cu(H_2O)_6]^{2+}$	Blue	+2	+2 (Ligand is neutral)
NH_3	$[Cu(NH_3)_4]^{2+}$	Dark blue	+2	+2 (Ligand is neutral)
Cl^-	$[CuCl_4]^{2-}$	(Yellow)	+2	–2 (Ligand has charge of –1)

Tip You need to remember the formulae of these complexes and the colours that are not in brackets.

Note the use of square brackets to surround a complex ion.

Bonding in complexes is mainly **dative covalent** bonding. The dative covalent bonding can be explained by the ligands having lone pairs of electrons and the transition metals having empty *d*-orbitals into which electrons can be donated.

Bidentate and polydentate ligands

Bidentate ligands have 'two pairs of teeth' and bind to the metal ion in two places. The example you need to know about is **ethanedioate**:

$$^{-}O-C(=O)-C(=O)-O^{-}$$

The lone pairs on the O^- ions are the places where the bidentate ligand attaches to the metal ion.

Polydentate means 'many teeth' and applies to ligands that can attach to the central metal ion in several places. The example you need to know about is **$EDTA^{4-}$**. The letters 'EDTA' represent the old name for the ion. Its structure is:

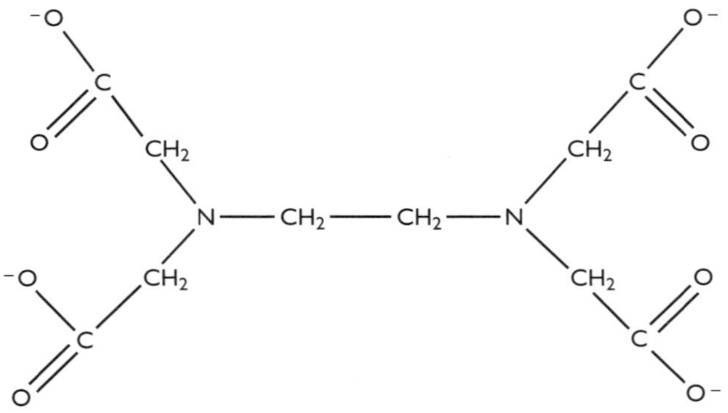

There are *six* places where this flexible ligand can attach to a central metal ion, the lone pairs on the two nitrogen atoms and the lone pairs on the four O^- ions.

Shapes of complexes

The shape is determined by the **coordination number** of the central atom or ion, i.e. *the number of bonds from the central ion to ligands*. The coordination numbers you need to know about are 4 and 6.

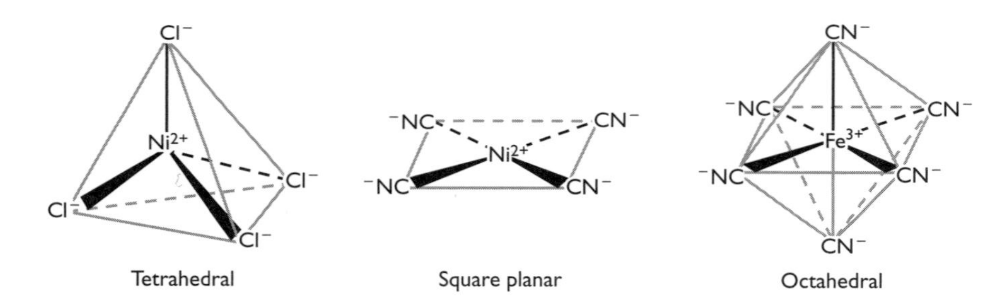

Tetrahedral | Square planar | Octahedral

Tip There are two arrangements for four ligands around a central ion. If asked just to draw a four-coordinate complex, you can choose either. You do not need to remember which metals form which four-coordinate complexes.

Note the definition of coordination number when polydentate ligands are used. Count the number of ion–ligand bonds, not the number of polydentate ligands. Thus $[Fe(ethanedioate)_3]^{3-}$ has a coordination number of *six*, not three.

Ligand exchange

This occurs when a transition metal ion swaps one set of ligands for another.

For example:

$$[Ni(H_2O)_6]^{2+} + 6NH_3 \rightleftharpoons [Ni(NH_3)_6]^{2+} + 6H_2O$$

Here the Ni^{2+} ion has exchanged six water molecules for six ammonia molecules.

Formation of coloured ions

The presence of ligands around the central ion affects the electrons in the *d*-subshell. The five *d*-orbitals are split into two groups at different energy levels. The energy difference between the split energy levels often corresponds to a frequency of visible light ($\Delta E = h\nu$).

Tip You need to know about this for Unit F335 but not for Unit F334.

What you need to know now is that:

- many transition metal compounds are coloured because of their unfilled *d*-subshells
- they absorb part of the visible spectrum of light
- the part they do not absorb (the complementary colour) is *transmitted* (*not emitted*) and this is the colour they appear

Some reactions with a colour change are given in the table below:

Ion	Reaction with NaOH(aq) and with small amounts of NH_3(aq)	Observation	Reaction with excess NH_3(aq)
Cu^{2+}	$Cu^{2+}(aq) + 2OH^-(aq) \rightarrow Cu(OH)_2(s)$	Light-blue precipitate	Precipitate dissolves to give $[Cu(NH_3)_4]^{2+}(aq)$ — dark-blue solution
Fe^{2+}	$Fe^{2+}(aq) + 2OH^-(aq) \rightarrow Fe(OH)_2(s)$	Dark-green precipitate	No change
Fe^{3+}	$Fe^{3+}(aq) + 3OH^-(aq) \rightarrow Fe(OH)_3(s)$	Rust-coloured precipitate	No change

Tip You need to learn these equations and colours.

The reactions are initially the same with both aqueous sodium hydroxide and aqueous ammonia, as the latter contains OH^- ions. Excess ammonia forms a complex with copper ions but not with iron(II) or iron(III) ions.

Catalytic behaviour

Many catalysts are either transition metals or their compounds.

They work as catalysts because:

- they have unfilled *d*-subshells. Thus, their surfaces can easily **adsorb** (form bonds with) gas molecules in **heterogeneous catalysis**. Examples are the use of iron in the Haber process and the use of platinum in catalytic converters.
- they have variable oxidation states. This enables them to work as **homogeneous catalysts** in aqueous solution. For example, if the reaction is $A + B \rightarrow C + D$, a faster route might be:

A + reduced X → C + oxidised X

B + oxidised X → D + reduced X

Either reduced X or oxidised X will act as a catalyst.

Redox

Revision

The chemical ideas here are from **Elements from the sea.**

Definitions

- Oxidation is *loss* of electrons.
- Reduction is *gain* of electrons.

Remember either **OIL RIG** (**o**xidation **i**s **l**oss, **r**eduction **i**s **g**ain) or **LEO the lion goes GER** (**l**oss of **e**lectrons is **o**xidation, **g**ain of **e**lectrons is **r**eduction).

Oxidation states

- The oxidation states of the atoms in a molecule (or neutral formula) add up to zero (this means that the oxidation state of an atom in an element is always zero).
- The oxidation states of the atoms in an ion add to the charge on the ion.
- There are certain fixed oxidation states for atoms in compounds:

Element	Oxidation state	Comments
H	+1	Except in H^-
O	–2	Except in peroxides (–1) and fluorine compounds
F	–1	
Cl	–1	Except in compounds with O or F
Group 1	+1	
Group 2	+2	

Examples

Species	Positive oxidation states	Negative oxidation states	Sum of oxidation states
SO_3	S +6	O 3 × –2	0
PCl_5	P +5	Cl 5 × –1	0
SO_4^{2-}	S +6	O 4 × –2	–2
NH_4^+	H 4 × +1	N –3	+1
N_2	N 0		0

Tip Don't forget the sign (even when positive) followed by the number.

- When the oxidation state of an atom *increases*, it has been *oxidised*.
- When the oxidation state of an atom *decreases*, it has been *reduced*.

Half-equations

These show the electrons being transferred. For example, the reaction between copper metal and silver ions involves the loss of two electrons from each copper atom and the gain of one electron by the silver ions.

$Cu(s) + 2Ag^+ \rightarrow Cu^{2+}(aq) + 2Ag(s)$

The half-equations are:

$Cu(s) \rightarrow Cu^{2+}(aq) + 2e^-$ (loss of electrons is *oxidation*)

$Ag^+(aq) + e^- \rightarrow Ag(s)$ (gain of electrons is *reduction*)

Other examples of half-equations are:

$2Cl^-(aq) \rightarrow Cl_2(aq) + 2e^-$ (*oxidation* of chloride to chlorine)

$MnO_4^-(aq) + 8H^+(aq) + 5e^- \rightarrow Mn^{2+}(aq) + 4H_2O(l)$

(*reduction* of manganate(VII) to manganese(II))

Note that the number of electrons transferred is equal to the oxidation state change each time. (Half-equations can be added together to make full equations — see p. 53.)

Cells and electrode potentials

You meet this topic in **The steel story**.

A piece of zinc is placed in some zinc(II) solution.

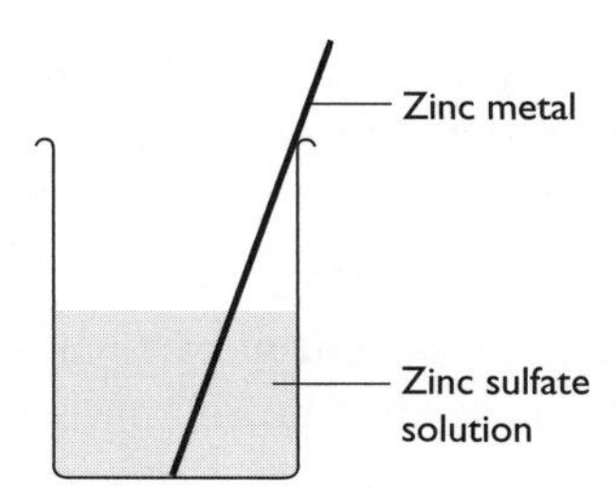

There is an exchange of electrons represented by the half-equation:

$Zn(s) \rightleftharpoons Zn^{2+}(aq) + 2e^-$

This is the equilibrium set up after a short while — the electrons going into the metal rod.

Another beaker contains a piece of copper in contact with copper(II) ions. An equilibrium is again set up:

$$Cu(s) \rightleftharpoons Cu^{2+}(aq) + 2e^-$$

Our knowledge of chemistry leads us to expect that the zinc equilibrium will lie further to the right than that for copper, since zinc is the more reactive metal. We can test this, by putting the two **half-cells** together to make a cell:

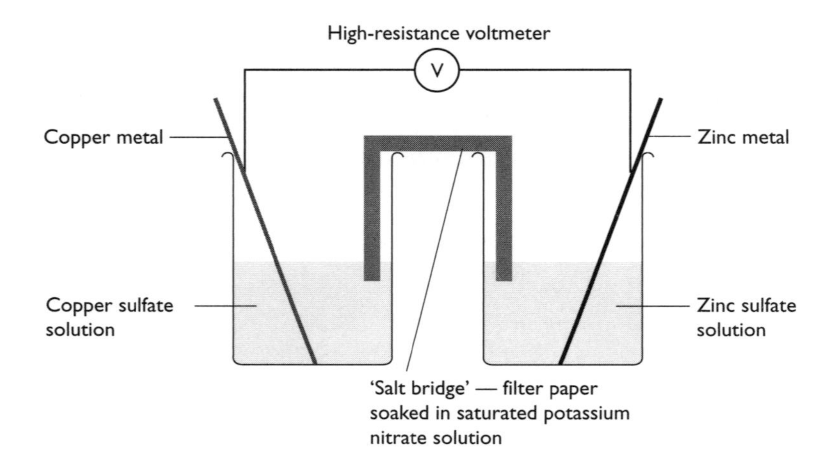

The zinc metal will become negative because the reaction $Zn(s) \rightarrow Zn^{2+} + 2e^-$ will occur to a greater extent than $Cu(s) \rightarrow Cu^{2+} + 2e^-$.

Standard conditions here are 298 K and 1.0 mol dm^{-3} solutions of the metal ions.

Thus the zinc rod is at a more negative potential than the copper rod. Electrons flow from the zinc to the copper in the external circuit. The voltmeter measures the potential difference between the two rods when very little current flows. Using 1.0 mol dm^{-3} solutions, the potential difference measured would be 1.1 V.

In order to measure the **electrode potential** of a half-cell, a standard electrode is needed. This is the hydrogen electrode. The cell needed to measure the **standard electrode potential** for a Zn/Zn^{2+} electrode is shown below.

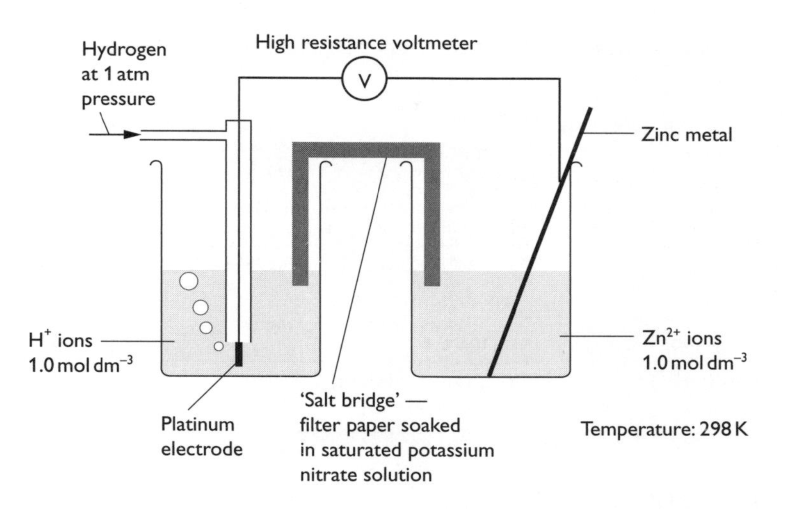

Tip You do not need to know the details of the hydrogen electrode.

The zinc electrode in this cell is the negative one and the meter reads 0.76 V. Thus we say that the *standard electrode potential* $E^\ominus$ for the Zn/Zn^{2+} electrode is –0.76 V.

Redox reactions between two species in solution

There are some half-cells in which the metal electrode is not one of the reagents. These have a platinum electrode and the standard state is when the concentrations are equal. Some examples are shown below.

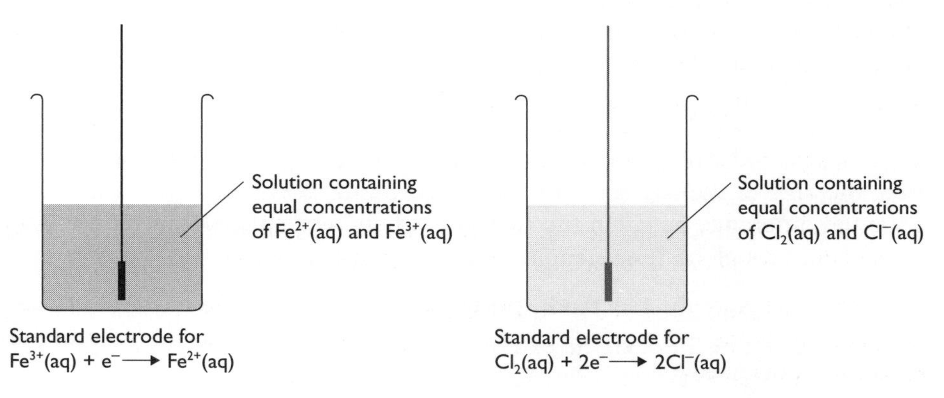

Some standard electrode potentials are given in the following table.

Half-cell	Half-equation	$E^{\ominus}$/V
$Zn^{2+}(aq)/Zn(s)$	$Zn^{2+}(aq) + 2e^- \rightarrow Zn(s)$	–0.76
$Fe^{2+}(aq)/Fe(s)$	$Fe^{2+}(aq) + 2e^- \rightarrow Fe(s)$	–0.44
$2H^+(aq)/H_2(g)$	$2H^+(aq) + 2e^- \rightarrow H_2(g)$	0 (by definition)
$O_2(g)/OH^-(aq)$	$O_2(g) + 2H_2O(l) + 4e^- \rightarrow 4OH^-(aq)$	+0.40
$I_2(aq)/2I^-(aq)$	$I_2(aq) + 2e^- \rightarrow 2I^-(aq)$	+0.54
$Fe^{3+}(aq)/Fe^{2+}(aq)$	$Fe^{3+}(aq) + e^- \rightarrow Fe^{2+}(aq)$	+0.77
$MnO_4^-(aq)/Mn^{2+}(aq)$	$MnO_4^-(aq) + 8H^+(aq) + 5e^- \rightarrow Mn^{2+}(aq) + 4H_2O(l)$	+1.51

Use of electrode potentials

Calculating $E^{\ominus}$ cell

This is done by taking the difference (*including* sign) between the two standard electrode potentials. The upper half-cell in the table is the negative terminal of the cell but the value of $E^{\ominus}_{cell}$ is quoted as a number without sign. The units are V. For example, $E^{\ominus}_{cell}$ for a cell consisting of a $Zn^{2+}(aq)/Zn(s)$ electrode and a $Fe^{3+}(aq)/Fe^{2+}(aq)$ electrode is 0.77 – (–0.76), which equals 1.53 V.

Making predictions about reacting substances

To do this, it is often useful to use a table such as the one above, arranged in order of $E^{\ominus}$ with the most negative at the top. Note that, in the table:

- the half-equations are always written with electrons on the left-hand side of the equation — therefore, the *oxidised form* (e.g. Zn^{2+} rather than Zn) on the *left*.
- the substance at the bottom left is the best *oxidising agent* and the substance at the top right is the best *reducing agent*
- when two electrodes are set up in a cell, *electrons will flow through the external circuit to the half-cell with the more positive electrode potential*

Thus if the $Zn^{2+}(aq)/Zn(s)$ and $I_2(aq)/2I^-(aq)$ half-cells are considered, the zinc one will supply electrons through the external circuit to the (more positive) iodine one. Therefore, the reactions that occur are:

$Zn(s) \rightarrow Zn^{2+}(aq) + 2e^-$ (i.e. the *reverse* of what is printed above) and

$I_2(aq) + 2e^- \rightarrow 2I^-(aq)$

This shows that zinc metal reduces iodine, but zinc(II) ions do *not* oxidise iodide ions. It can also be seen that zinc metal will react with any of the substances on the left-hand side of the half-equations below –0.76 V, i.e. Fe^{2+}, H^+, Fe^{3+} and MnO_4^-/H^+. However, I^- will react only with any of the substances on the left-hand side of the half-equations below +0.54 V, i.e. MnO_4^-/H^+ and Fe^{3+}.

To write equations for feasible reactions

This follows on from the above. When reaction between two reagents is predicted, we say that the reaction between them is *feasible*. (It may not happen if the activation energy is too large and thus the reaction is very slow.)

The steps that are needed to write equations are as follows:

- Write down the upper half-equation reversed (i.e. electrons on the right).
- Copy the lower half-equation as it is written.
- Make equal the number of electrons transferred in each half-equation by suitable multiplication.
- Add the equations together (the number of electrons will then cancel).

Example 1

Zinc will react with Fe^{3+}. The half-equations are:

$Zn(s) \rightarrow Zn^{2+}(aq) + 2e^-$

(upper half-equation reversed)

$2Fe^{3+}(aq) + 2e^- \rightarrow 2Fe^{2+}(aq)$

(lower equation doubled to get $2e^-$)

Adding: $Zn(s) + 2Fe^{3+}(aq) \rightarrow 2Fe^{2+}(aq) + Zn^{2+}(aq)$

(the numbers of electrons cancel out)

Example 2

Purple manganate(VII) is reduced in acid conditions by iodide ions, $I^-(aq)$. The half-equations are:

$10I^-(aq) \rightarrow 5I_2(aq) + 10e^-$

(upper half-equation reversed and multiplied by 5 to get 10 electrons)

$2MnO_4^-(aq) + 16H^+(aq) + 10e^- \rightarrow 2Mn^{2+}(aq) + 8H_2O(l)$

(lower equation doubled to get 10 electrons)

Adding: $2MnO_4^-(aq) + 16H^+(aq) + 10I^-(aq) \rightarrow 5I_2(aq) + 8H_2O(l) + 2Mn^{2+}(aq)$

Rusting

This is an electrochemical process. The two half-equations are:

$Fe^{2+}(aq) + 2e^- \rightarrow Fe(s)$ $\quad E^\ominus = -0.44\ V$

$O_2(g) + 2H_2O(l) + 4e^- \rightarrow 4OH^-(aq)$ $\quad E^\ominus = +0.40\ V$

The overall equation is derived:

Equation 1: $Fe(s) \rightarrow Fe^{2+}(aq) + 2e^-$ (upper equation reversed)

Equation 2: $\frac{1}{2}O_2(g) + H_2O(l) + 2e^- \rightarrow 2OH^-(aq)$ (lower equation halved to get $2e^-$)

Adding: $\frac{1}{2}O_2(g) + H_2O(l) + Fe(s) \rightarrow Fe^{2+}(aq) + 2OH^-(aq)$

The next reactions are:

$Fe^{2+}(aq) + 2OH^-(aq) \rightarrow Fe(OH)_2$

$Fe(OH)_2$ is then oxidised by oxygen in the air to $Fe_2O_3.xH_2O$ (rust)

Equation 2 happens where there is *more* oxygen. Equation 1 (the corrosion of iron) happens where there is *less* oxygen — often at the bottom of a pit in the metal. This causes serious damage.

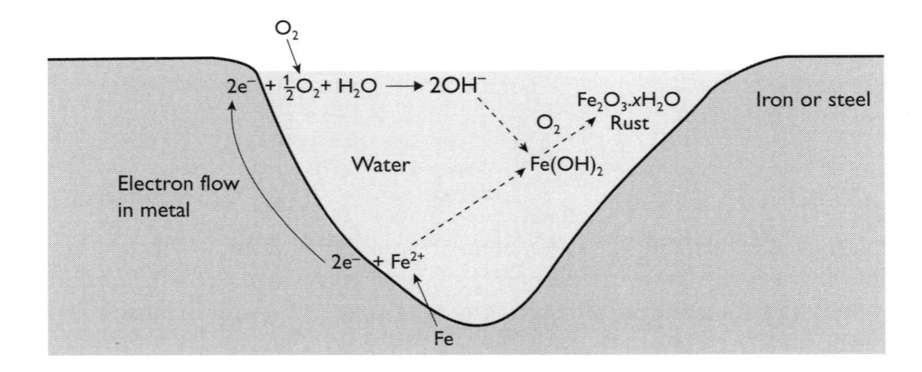

Tip It would be useful to learn this diagram.

Rusting can be inhibited by:

- painting, oiling, greasing or covering with a polymer to exclude oxygen and water (but iron rusts more at any breaks in the coating — just like it does at the bottom of a pit, as there is less oxygen there than at the surface)
- plating with a metal higher in the electrode potential series (e.g. zinc), which will corrode instead of the iron if the zinc is scratched. Plating with zinc is called **galvanising**. Another way to protect an iron object is to weld zinc blocks to it. A cell is set up in which the zinc corrodes (see below).

Note: tin plating is used extensively, but tin is lower in the electrode potential series, so it works, like paint, by covering the steel.

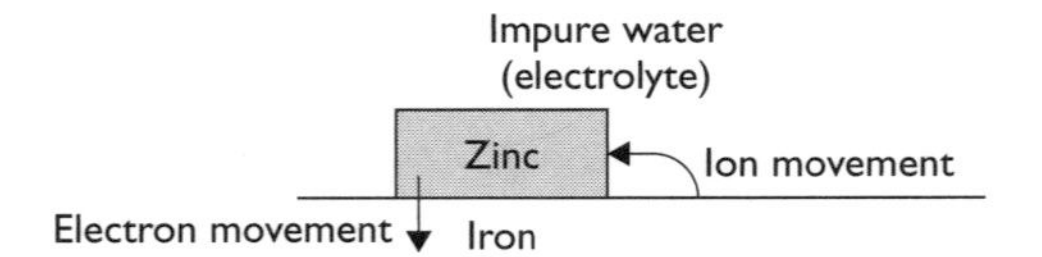

Analytical methods

Colorimetry

You meet the colorimeter in **The steel story**. A colorimeter is used to measure the concentration of coloured substance in a solution. It works by measuring the amount of light of a particular colour (determined by a coloured filter) that passes through a solution. A spectrophotometer enables the wavelength of the complementary colour to be chosen and used.

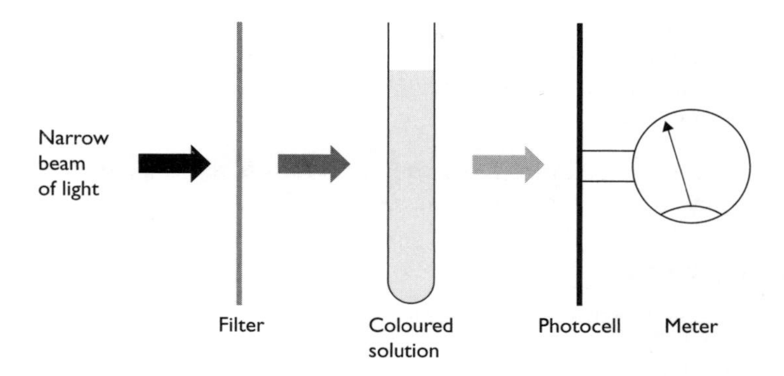

The colour of the filter is the complementary colour to the colour of the solution. For example, a green filter is used for a purple solution, because purple solutions absorb green light and transmit purple light. The meter reads the **absorbance** (% light absorbed).

The best way to relate absorbance to concentration is to use a **calibration curve**.

The steps in measuring the concentration of a coloured solution with a colorimeter are:

- Select a suitable filter (complementary colour) for the colorimeter.
- Adjust absorbance to zero using a 'blank' tube containing water.
- Make up several standard solutions of the coloured substance. ('Standard' means *known concentration.*)
- Measure the absorbance for each of these concentrations.
- Plot a graph of concentration against absorbance. This is the **calibration curve**.
- Find the absorbance for the solution of unknown concentration.
- Read off the concentration of the unknown solution from the calibration curve.

Manganate(VII) titrations and mole calculations

You meet these in **The steel story**.

Manganate(VII) titrations can be used to measure the concentration of a substance in aqueous solution if it reduces manganate(VII) ions in acid conditions. As you can see from the table on p. 52, manganate(VII) is easily reduced, so it is suitable to titrate with many reducing agents.

The steps in carrying out such a titration are as follows:

- Fill a burette with a potassium manganate(VII) solution of *known concentration.*
- Add a *known volume* (measured by *volumetric pipette*) of substance X to a conical flask.
- Add an approximately equal volume of 1 mol dm^{-3} sulfuric acid to the flask.
- Titrate the solution of substance X with manganate(VII), while swirling the flask, until the *first permanent trace of pink colour* appears.
- Repeat until several titration results are within 0.1 cm^3 and take their average.

To calculate the concentration of substance X:

- Use half-equations to write the equation for the reaction of X with MnO_4^- (see p 53).
- Use your titration result and the known concentration of the manganate(VII) solution to work out moles of manganate(VII) added.
- Use the equation to work out the amount in moles of X in the flask.
- Use the known volume of X and the amount in moles of X to work out the concentration of X.

You may also be expected to work out concentrations of reagents in other titrations, such as acid–base ones. Follow the same steps to do this.

Mass spectrometry

You meet this topic in **What's in a medicine?**

Mass spectrometry tells us the M_r of the molecule and gives information about its structure from the fragments formed in the spectrometer. The mass spectrum of ethanol is shown below.

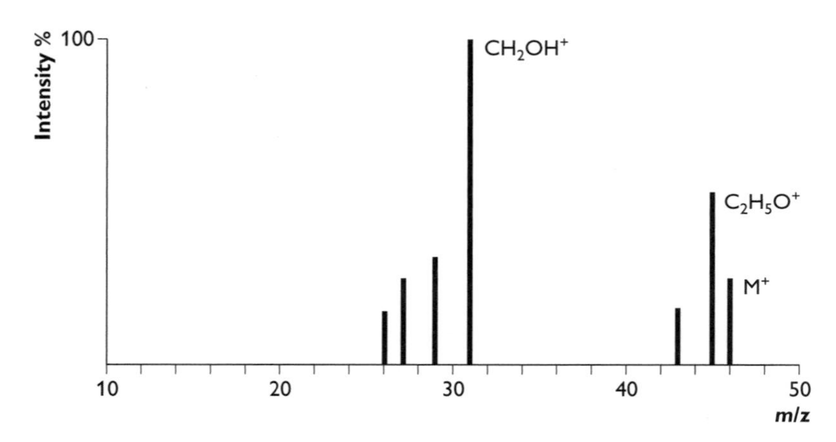

Note that all the peaks are caused by singly charged positive ions. You *must* show this + sign if identifying any peak. The peak of highest mass (called the molecular ion peak or M^+) gives the M_r — here 46. The other peaks are fragments. A good way to begin to analyse fragments is to look at the *masses lost*. A loss of 1 (45) indicates a loss of a single hydrogen, almost certainly the hydroxyl one. A loss of 15 (31) indicates loss of CH_3. Notice there is no peak at 15, which means that the methyl groups are formed as radicals rather than ions. In aromatic structures, a peak at 77 indicates the $C_6H_5^+$ ion.

Tip You need to remember the significance of losses of 15 and 77 in a mass spectrum.

Infrared spectroscopy

You met this topic in **The polymer revolution**.

Infrared spectroscopy tells us some of the functional groups found in molecules. The infrared spectrum of ethanol (liquid film) is shown on the following page.

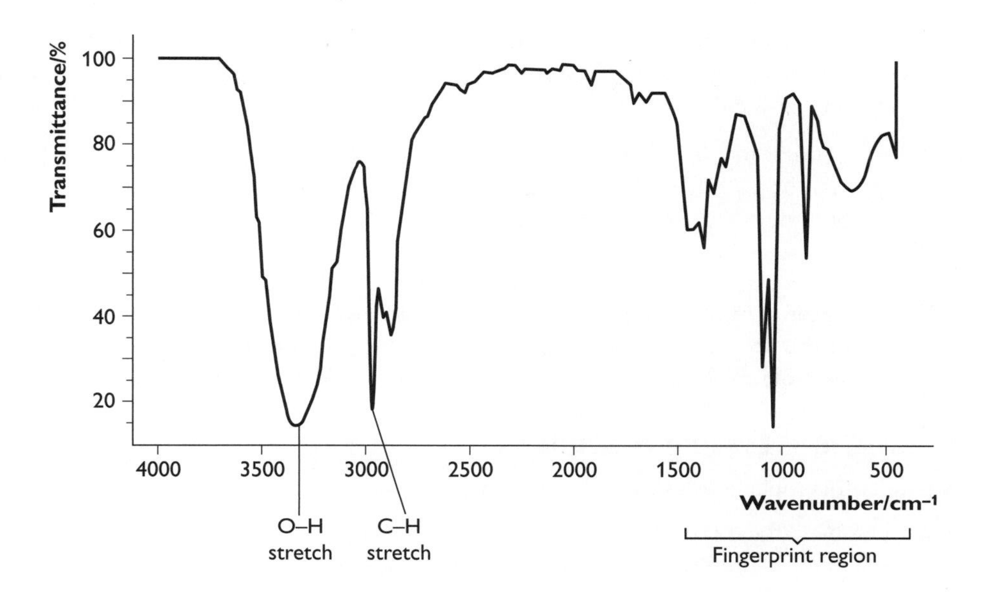

The O–H and C–H stretches show up well. C–O is also present at 1050 cm^{-1}, but it is in the 'fingerprint region', which is difficult to interpret since other stretching (and bending) modes can get in the way. Usually, it is best to look at absorptions above 1500 cm^{-1}, which are much clearer. As well as the two stretches shown, always look out for C=O around 1700 cm^{-1}. The *Data Sheets* enable you to make a judgement about which type of C=O compound is present (e.g. aldehydes are usually in the range 1720–1740 cm^{-1}, while esters are in the range 1735–1750 cm^{-1}).

Tip Don't forget the units (cm^{-1}) if identifying any peaks.

Industrial and green chemistry

Extraction and purification of metals

Most metals exist in nature as their compounds; only very unreactive metals such as gold are found *native*. This means that the compounds have to be *reduced* to yield the metal. The metals highest in the reactivity series are produced by **electrolysis** — for example, the extraction of aluminium, where molten aluminium oxide is electrolysed and aluminium forms at the cathode:

$$Al^{3+} + 3e^- \rightarrow Al$$

Carbon reduction is used for metals in the middle of the reactivity series — for example, iron, where the reducing agent is mainly carbon monoxide:

$$Fe_2O_3 + 3CO \rightarrow 2Fe + 3CO_2$$

Impurities in the ore are sometimes removed at the same time — for example, SiO_2 is removed by calcium oxide in the blast furnace:

$CaO + SiO_2 \rightarrow CaSiO_3$

Metals are sometimes purified by electrolysis — for example, copper. The impure copper forms the cathode. Pure copper plates onto the cathode and the impurities fall to the bottom of the electrolysis cell.

$Cu^{2+}(aq) + 2e^- \rightarrow Cu(s)$

Tip You need to be able to explain these chemical processes, given the necessary information. Thus, you do not need to remember any of the above reactions or methods, but they do illustrate some possible contexts that you may be asked to comment on.

Making and testing medicines

You meet this topic in **What's in a medicine?**

When chemists have found an effective medicine, they try to improve it by varying its chemical structure. The structure is slightly varied and the medicine is tested to see if it is more effective. This testing investigates whether lower doses can be used effectively, what size of dose is harmful, whether the medicine can be given in a different form (e.g. liquid rather than solid), what the side effects are and how these can be reduced.

The technique for making many compounds of similar but slightly different chemical constitution is called **combinatorial chemistry**. For example, if a medicine is an ester, many related esters are made in small tubes by an automated process. The compounds are evaluated by large-scale screening.

When medicines are tested, the last step is the **clinical trial**, when volunteers take the medicine under carefully controlled conditions. These trials are designed to answer three crucial questions:

- Is it safe? (Are side-effects small? Is the effective dose close to an overdose?)
- Does it work? (Does it work in small doses? How specific is it?)
- Is it better than the standard treatment? (More specific, smaller doses, fewer side-effects?)

Tip You need to remember the three bullet points, but you need only to understand the follow-up questions in brackets.

Atom economy and reaction type

You meet this topic in **What's in a medicine?**

Atom economy is defined as:

$$\frac{M_r \text{ of useful products}}{M_r \text{ of the reactants used}} \times 100$$

Note that this is the same as:

$$\frac{M_r \text{ of useful products}}{M_r \text{ of all products}} \times 100$$

This formula can be more useful.

Manufacturers design modern processes with as high an atom economy as possible, to avoid wasting material. Wasted material is inefficient and has to be carefully (and expensively) disposed of to avoid pollution of the environment.

Manufacturers achieve higher atom economy by trying to move away from reactions such as substitution and elimination, which have low atom economy, towards addition and rearrangement, which have 100% atom economy.

Reaction type	Summary	Example
Rearrangement	A → B	Isomerisation of an alkane $CH_3CH_2CH_2CH_2CH_3 \rightarrow CH_3CH(CH_3)CH_2CH_3$ atom economy 100%
Addition	A + B → C	Addition of bromine to an alkene $CH_2CH_2 + Br_2 \rightarrow CH_2BrCH_2Br$ atom economy 100%
Condensation (addition + elimination)	A + B → C + H_2O or NH_3 or HCl	Esterification $C_2H_5OH + CH_3COOH \rightarrow C_2H_5OOCCH_3 + H_2O$ atom economy = 88 × 100/106 = 83.0%
Substitution	A + B → C + D	Reaction of an alcohol with HCl $CH_3OH + HCl \rightarrow CH_3Cl + H_2O$ atom economy = 50.5 × 100/68.5 = 73.7%
Elimination	A → B + C	Alcohol to alkene $C_2H_5OH \rightarrow CH_2CH_2 + H_2O$ atom economy = 28 × 100/46 = 60.9%

The order is always: addition and rearrangement (both 100%) > condensation and substitution > elimination

Tip You need to know this order and the summary equations in the table. You should be able to recall examples using organic compounds and you should be able to calculate the atom economies.

The polymer 'life cycle'

You meet this topic in **The materials revolution**.

Manufacture of polymers

Methods have to be devised to ensure that the raw materials are produced in a way that:

- minimises hazardous waste
- uses the minimum amount of energy
- releases the least amounts of carbon dioxide and other polluting gases

Then the polymerisation process itself must also be designed to make as little impact on the environment as possible.

Recycling polymers

Landfill

Until recently, most polymers when thrown away went to landfill. Disposing of any material in this way wastes land resources, but polymers are particularly bad because they take a long time to degrade. Biodegradable polymers can help alleviate this problem.

Burning

This produces energy but also polluting gases (e.g. CO_2 and HCl). It is only an option in special incinerators where the waste products can be trapped.

Depolymerising

If a polymer can be converted back into its monomers (depolymerised), it can then be repolymerised. This is possible for some materials, but often plastics are mixed in with other waste and separating them is costly and time-consuming.

Cracking

This is a modern method of breaking down polymers into smaller molecules that can be used as chemical feedstocks.

Recycling steel

You meet this topic in **The steel story**.

- Steel packaging can easily be recycled. It can be separated from other metals by its magnetic properties. (Aerosol containers need special treatment because of the flammable gases they contain.)
- The steel is always melted for re-use and this incineration removes any dirt or other organic material.
- Steel is added to a blast furnace where it not only replaces some iron ore but also controls the temperature of the furnace. (One steel-making method, the electric arc process, runs entirely on scrap steel.)

Enzymes in industry

You meet this topic in **The thread of life**. You do not have to learn any examples here. You simply need to be able to interpret examples or talk about the subject in the general terms given below.

Enzymes are useful in industrial processes because:

- they are specific catalysts, thus a reaction can be chosen with high atom economy
- they often work at a lower temperature than other catalysts (thus they use less energy)
- they often require milder conditions (e.g. no concentrated acids)
- the solvent and co-products are often less toxic in enzyme-catalysed reactions

Questions & Answers

In this section of the guide there are five questions that between them test every lettered statement in the specification and a selection of synoptic material as well. They represent the kinds of question you will get in the unit test, in that they start with a context and they contain a wide range of subject matter. Unlike the real unit test, there are no lines or spaces left for your answers. Instead, the presence of a space or a number of lines is indicated. The number of marks is also shown. However, taken together, these questions are much longer than a single paper, so do not try to do them all in 90 minutes.

After each question, you will find the answers of two candidates — Candidate A and Candidate B (using different candidates for each question). In each case, Candidate A is performing at the C/D level, while Candidate B is an A-grade candidate.

Examiner's comments

All candidate responses are followed by examiner's comments. These are preceded by the icon *e* and indicate where credit is due. In the weaker answers, they also point out areas for improvement, specific problems and common errors.

How to use this section

- Do the question, giving yourself a time limit of a minute a mark; do not look at the candidates' answers or examiner's comments before you attempt the question yourself.
- Compare your answers with the candidates' answers and decide what the correct answer is; still do not look at the examiner's comments while doing this.
- Finally, look at the examiner's comments.

Completing this section will teach you a lot of chemistry and vastly improve your exam technique.

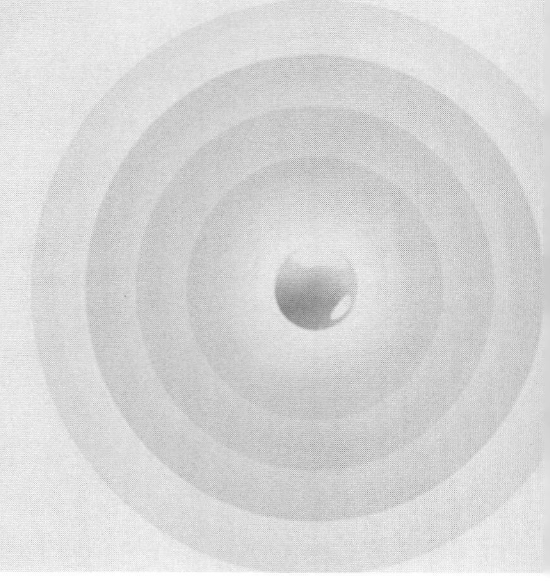

Question 1

Stainless steel

A steel plant sets out to make stainless steel for surgeons' knives. The steel is made from blast furnace iron. Elements such as carbon and sulfur have to be removed, while more manganese and chromium are added.

(a) Suggest a reason why blast furnace iron contains carbon. ***(2 lines)*** (1 mark)

(b) Sulfur is the first element to be removed during steelmaking by reaction with magnesium. Write a balanced chemical equation for this reaction. ***(space)*** (1 mark)

(c) Iron scrap is added to the blast furnace iron before the steel is made. Give *one* reason for doing this. ***(2 lines)*** (1 mark)

(d) Manganese is an element that has to be added at the end of the steelmaking process.

(i) Use your periodic table to complete the electronic structure of the Mn^{2+} ion. (1 mark)

1s	2s	2p	3s	3p	3d	4s
↑↓	↑↓	↑↓ ↑↓ ↑↓	↑↓	↑↓ ↑↓ ↑↓	☐☐☐☐☐	☐

(ii) Explain, in terms of your answer to (i), why manganese is described as a transition metal. ***(2 lines)*** (2 marks)

(e) The manganese in steel is oxidised to the purple MnO_4^- ion, which can be used to determine the percentage of manganese in the steel.

(i) Give the oxidation states of manganese in Mn^{2+} and MnO_4^-. ***(1 line)*** (2 marks)

(ii) Give the *general property* of transition metals that is illustrated by part (i). ***(1 line)*** (1 mark)

(iii) Explain this property in terms of electron energy levels. ***(2 lines)*** (2 marks)

(iv) Explain, in terms of parts of the visible spectrum, why an MnO_4^- solution looks purple. ***(3 lines)*** (2 marks)

(v) Describe how a colorimeter can be used to measure the concentration of a MnO_4^- solution that is approximately 0.001 mol dm^{-3}. ***In your answer you should use technical terms, spelt correctly (5 lines)*** (5 marks)

(f) Potassium manganate(VII) titrations can be used to calculate the percentage of iron in substances such as iron tablets. In an experiment, an iron tablet of mass 1.60 g is dissolved in dilute sulfuric acid and the solution is made up to 250 cm^3. 10.0 cm^3 portions of the solution are titrated against 0.0100 mol dm^{-3} manganate(VII) solution. 21.1 cm^3 are required.

(i) Name the piece of apparatus that is used to:

- **make the solution up to 250 cm^3** ***(1 line)***
- **measure out the 10.0 cm^3 of solution** ***(1 line)*** (2 marks)

(ii) What would you see at the end-point of the titration? *(1 line)* (1 mark)

(iii) The equation for the reaction of the Fe^{2+} from the iron tablet with manganate(VII) is:

$MnO_4^- + 5Fe^{2+} + 8H^+ \rightarrow Mn^{2+} + 5Fe^{3+} + 4H_2O$

Calculate the percentage of iron in the iron tablet. *(space)* (5 marks)

(iv) Iron(III) salts react with NaOH in aqueous solution. Write an ionic equation for this reaction and describe what is seen when it occurs. *(space)* (3 marks)

(g) The chromium in the steel can be identified by the complexes it forms. Two such complexes are shown in the equilibrium below.

$[Cr(NH_3)_5Cl]^{2+}(aq) + NH_3(aq) \rightleftharpoons [Cr(NH_3)_6]^{3+}(aq) + Cl^-(aq)$

Complex A **Complex B**

(i) Complex A has two ligands, one of which is charged. Give the formulae of the two ligands. *(1 line)* (2 marks)

(ii) Draw a diagram to illustrate the shape of complex A and describe the shape. *(space)* (2 marks)

(iii) Name the *type* of reaction in the equation above. *(1 line)* (1 mark)

(iv) $EDTA^{4-}$ is a *polydentate* ligand which forms a stable complex with chromium ions. Explain what 'polydentate' means. *(2 lines)* (1 mark)

(h) Iron is used as a catalyst in the Haber process. Explain, in terms of their electronic structure, why transition metals often make good catalysts for gas reactions. *(4 lines)* (3 marks)

Total: 38 marks

Candidates' answers to Question 1

Candidate A

(a) It comes from the iron ore.

Candidate B

(a) It is added to reduce the iron ore.

Candidate A is wrong and does not score. There is little carbon in iron ore and the main impurity is SiO_2. Candidate B is correct. Carbon is added to form the carbon monoxide needed to reduce the iron oxide ore to iron in a blast furnace.

Candidate A

(b) $Mg + 2S \rightarrow MgS_2$

Candidate B

(b) $Mg + S \rightarrow MgS$

Candidate A is wrong because he/she does not know that the charge on the sulfide ion is 2–. This can be worked out from the position of sulfur in the periodic table. Candidate B is correct and scores the mark.

Candidate A

(c) To adjust the temperature of the furnace

Candidate B

(c) To allow the recycling of iron

Both are acceptable answers, so they each score 1 mark.

Candidate A

(d) (i)

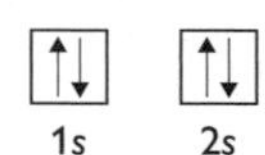
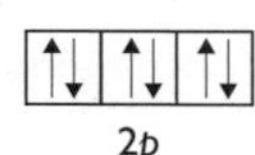
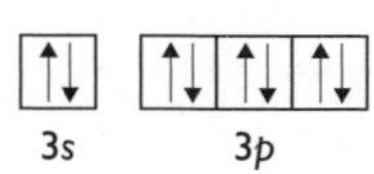
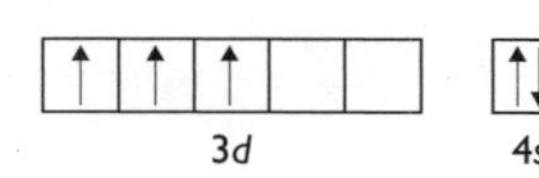

Candidate B

(d) (i)

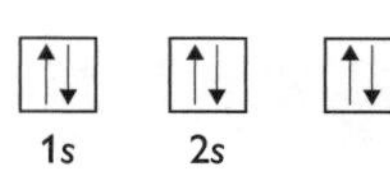
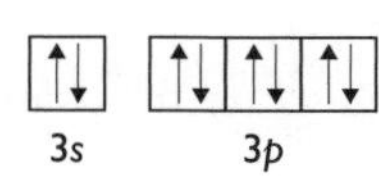
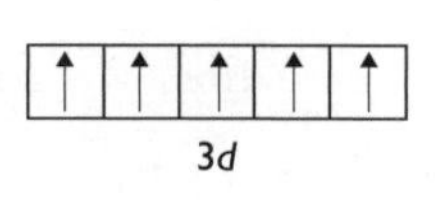
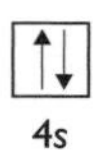

Neither candidate has done particularly well here. Candidate A has not appreciated that the two electrons manganese loses come from the 4s-subshell. They may be first in but they are also first out. Candidate B has written the electron structure for the *atom*, not the *ion*. Neither would score the mark.

Candidate A

(d) (ii) It is in the *d*-block of the periodic table.

Candidate B

(d) (ii) Some of its *d*-subshells are incomplete.

Candidate A's answer is wrong. Candidate B's answer is nearer but still only scores 1 out of 2 because it does not say that *in some of its compounds* the d-subshell is incomplete.

Candidate A

(e) (i) +2; +8

Candidate B

(e) (i) +2; +7

Candidate B is correct and scores both marks. Candidate A has perhaps forgotten that the oxidation states of atoms in an ion must add to the *charge on the ion* — in this case −1. Candidate A therefore scores 1 mark for +2.

Candidate A

(e) (ii) Different charges on their ions

Candidate B

(e) (ii) They have variable oxidation states.

Candidate A is not quite right. The ion charges may vary, but it is the variation of oxidation state that is important. Candidate B is correct, for 1 mark.

Candidate A

(e) (iii) Many different stable arrangements

Candidate B

(e) (iii) There are many possible stable arrangements of the 3*d*-electrons.

Candidate A has offered part of the answer and scores 1 mark. Candidate B has included the extra detail and gains the second mark.

Candidate A

(e) (iv) It absorbs in the green region, so it looks purple.

Candidate B

(e) (iv) It absorbs the complementary colour to purple.

Both candidates have been rather vague and only score 1 out of 2 marks. The marks are for: *absorbs green (or complementary colour); transmits purple.* Thus, neither candidate has given sufficient detail to score the second mark.

Candidate A

(e) (v) Put tube of water into the colorimeter and calibrate it. Then put in coloured tube. Read meter. Meter reading gives concentration.

Candidate B

(e) (v) Make up a series of solutions of manganate(VII) in the right range. Measure the colorimeter readings for each of these solutions and plot a calibration graph. Then take the reading for the unknown solution and read its concentration off the graph.

Candidate A scores 1 mark for the idea of zeroing the colorimeter and another mark for 'put in coloured tube...read meter', so he/she earns a total of 2 marks only. Candidate B misses 1 mark for not mentioning a filter and not zeroing the colorimeter (these are alternative ways of scoring the first mark) and misses another by not saying that the manganate(VII) solutions are of known concentration. Candidate B then loses the mark for quality of written communication by not using the word 'absorbance' instead of 'colorimeter reading'. Candidate B scores 2 marks for taking the reading of the unknown and reading its concentration, so earns 2 out of 5 overall.

Candidate A

(f) (i) Measuring cylinder; burette

Candidate B

(f) (i) Standard flask; pipette

Candidate A has forgotten the AS practical work. Measuring cylinders are not accurate enough for this type of work and, while a burette could be used to measure 10.0 cm^3 of solution, a chemist would always use a pipette in this context. So, no marks here. Candidate B has named the right equipment and scores both marks. Note that a standard flask is sometimes called a graduated or volumetric flask. Any of these would be acceptable.

Candidate A

(f) (ii) The indicator changes colour.

Candidate B

(f) (ii) A pale pink colour is visible.

Candidate A has confused manganate(VII) and acid–base titrations. Manganate titrations are self-indicating and the colour change should be given so Candidate A scores no marks. The best answer is the *first permanent trace of pink (or purple) colour*. Candidate B just scores for the answer here.

Candidate A

(f) (iii) $\% = 25 \times 2.11 \times 10^{-4} \times 55.8/1.6 = 0.18\%$

Candidate B

(f) (iii) moles MnO_4^- used in titration $= 2.11 \times 10^{-4}$ mol
moles Fe in one $10\ cm^3$ pipette $= 5 \times 2.11 \times 10^{-4}$ mol
moles Fe in $250\ cm^3$ flask $= 5 \times 25 \times 2.11 \times 10^{-4}$ mol
mass Fe in $250\ cm^3$ flask $= 56 \times 5 \times 25 \times 2.11 \times 10^{-4}$ mol $= 1.4770$ g
percentage $= (1.4770/1.60) \times 100 = 92.3\%$

Candidate B has written out a careful calculation and scores 5 marks (1 for each line). Candidate A has tried to do the whole calculation at once and probably got confused in the process. However, the examiner can see that there are two steps missing: the factor of 5 that comes from the equation and the 100 to turn the final answer into a percentage. So candidate A scores 3 marks by 'error carried forward'.

Candidate A

(f) (iv) An orange precipitate
$Fe_2(SO_4)_3 + 3NaOH \rightarrow Fe(OH)_3 + 3Na_2SO_4$

Candidate B

(f) (iv) A green precipitate
$Fe^{2+} + 2OH^- \rightarrow Fe(OH)_2$

Candidate A has given the correct colour of the precipitate and the correct formula, so he/she scores 2 marks. The equation is unbalanced and not ionic anyway. Candidate B has misread the question and answered for iron(II) not iron(III). Candidate B scores 1 mark for a balanced ionic equation.

Candidate A

(g) (i) NH_3; Cl^-

Candidate B

(g) (i) Ammonia and chlorine

Candidate A scores both the marks for correct formulae. Candidate B has not read the question properly and has given names not formulae. 'Ammonia' is acceptable here but not 'chlorine' as the formula, with charge, of the chloride ion was needed. Candidate B scores 1 mark.

Candidate A

(g) (ii) Octahedral

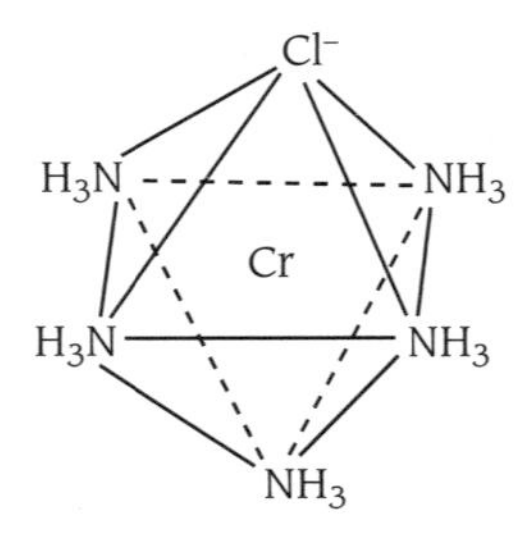

Candidate B

(g) (ii) Six-coordinate

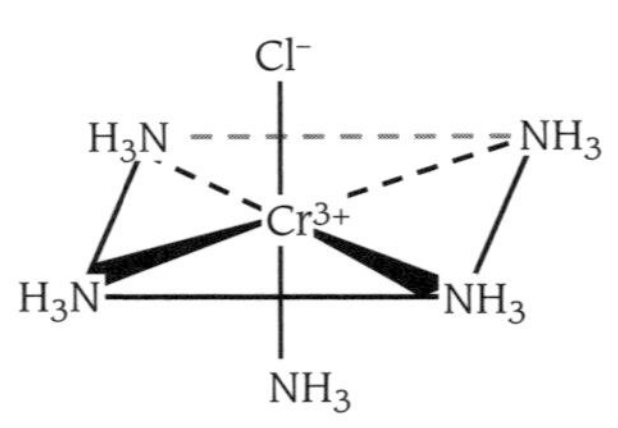

Here are two rather different answers for the diagram. However, both represent the structure correctly, so they each score the mark. Candidate A has correctly described the shape and correctly scored the second mark. Candidate B has written good chemistry, but it does not answer the question, so does not score the second mark.

Candidate A

(g) (iii) Ligand exchange

Candidate B

(g) (iii) Oxidation

Candidate A is correct and scores the mark. Candidate B is incorrect. There is no oxidation occurring here. Chromium has the oxidation state +3 in both complexes.

Candidate A

(g) (iv) It has lots of teeth.

Candidate B

(g) (iv) It has many (six) points of attachment to the central ion.

Candidate A has used the phrase he or she has learnt to remember the meaning of polydentate. To score, it was necessary to interpret what 'teeth' meant. Candidate B does this and gains the mark.

Candidate A

(h) Catalysts speed up chemical reaction but are unchanged at the end. Iron does this well because of its incomplete *d*-subshell.

Candidate B

(h) Transition metals have unfilled *d*-orbitals. This enables gas molecules to be adsorbed (chemically bonded) to the surface, where reactions can more easily occur.

Candidate A starts vaguely but does score 1 mark for mentioning the incomplete *d*-subshell. Candidate B also mentions this (using different but equally acceptable terminology). The next sentence scores the second mark. However, the last sentence is just too vague to score. Some mention of lowering the activation enthalpy was needed for this marking point. Candidate B scores 2 out of 3 marks.

Question 2

The enzyme urease

The enzyme *urease* is present in many simple organisms. It catalyses the hydrolysis of toxic urea into ammonia and carbon dioxide.
The rate of this reaction can be followed in the laboratory by measuring the amount of alkaline ammonia produced.

(a) Suggest a method of measuring the amount of ammonia produced. *(3 lines)* (2 marks)

(b) Some data for the laboratory reactions of urease are given in the table below.

Experiment	Concentration of enzyme/mol dm^{-3}	Concentration of urea/mol dm^{-3}	Rate of reaction/mol dm^{-3} s^{-1}
A	1.0×10^{-5}	0.01	1.0×10^{-3}
B	2.0×10^{-5}	0.01	2.0×10^{-3}
C	1.0×10^{-5}	0.02	2.0×10^{-3}
D	1.0×10^{-5}	0.10	5.0×10^{-3}
E	1.0×10^{-5}	0.20	5.0×10^{-3}

(i) Use the results of experiments A, B and C to determine (at low substrate concentration) the *orders* of reaction with respect to enzyme concentration, [E], and substrate concentration, [S]. *(2 lines)* (2 marks)

(ii) Write a *rate equation* in terms of [E] and [S]. *(1 line)* (2 marks)

(iii) Write down the *overall order* of the reaction. *(1 line)* (1 mark)

(iv) Give the units of the *rate constant* from your rate equation. *(space)* (2 marks)

(v) Sketch the graph you would expect for *substrate concentration* (y-axis) against *time* (x-axis) when the substrate concentration has the order you determined in part (i). Mark two successive half-lives on your sketch. *(space)* (2 marks)

(vi) Use the results of experiments D and E to show that the order of reaction with respect to substrate at high substrate concentration is zero. *(2 lines)* (1 mark)

(vii) Explain why the order of an enzyme reaction with respect to the substrate is zero at high substrate concentration. *(3 lines)* (2 marks)

(c) Some students tried using compound A as a substrate for urease. No reaction occurred.

(i) Name the functional group in compound A. *(1 line)* (1 mark)

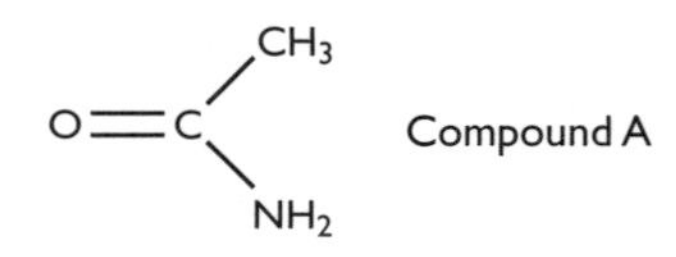

(ii) Write the equation for the hydrolysis of compound A with water, which occurs very slowly. *(space)* (2 marks)

(iii) **Suggest, in terms of enzyme structure, why urease would not catalyse the hydrolysis of compound A.** ***In your answer you should make it clear how the steps you describe are linked together. (5 lines)*** (5 marks)

(iv) **Explain, in terms of enzyme structure, what is meant by an 'enzyme inhibitor'.** ***(3 lines)*** (2 marks)

(v) **Describe *one* other condition (other than the use of inhibitors) that stops an enzyme working and describe how it does this.** ***(3 lines)*** (3 marks)

(vi) **Give *one* reason why the use of enzymes in industry is better for the environment.** ***(2 lines)*** (1 mark)

(d) Some students hydrolysed a sample of urease. Glycine and alanine were among the amino acids in the product.

H_2N-CH_2-COOH

Glycine

$H_2N-CH(CH_3)-COOH$

Alanine

(i) **Suggest a technique that could have been used to separate the amino acids prior to identification.** ***(1 line)*** (1 mark)

(ii) **Draw the structure of the zwitterion that glycine forms in solution.** ***(space)*** (2 marks)

(iii) **Draw the structure of *one* dipeptide formed between a molecule of glycine and a molecule of alanine.** ***(space)*** (2 marks)

(iv) **What *type* of reaction occurs when the molecule in part (iii) is formed from the two amino acids?** ***(1 line)*** (1 mark)

(v) **What name is given to the link between the two amino acids in the dipeptide?** ***(1 line)*** (1 mark)

(vi) **One of the two amino acids is *chiral*. Draw the two stereoisomers that result from this property.** ***(space)*** (2 marks)

(e) (i) **There are three levels of protein structure. Briefly describe each.**

- **Primary** ***(2 lines)***
- **Secondary** ***(2 lines)***
- **Tertiary** ***(2 lines)*** (3 marks)

(ii) **Explain why the primary structure of an enzyme is important in the function of the enzyme.** ***(2 lines)*** (2 marks)

(iii) **Give *two* interactions that determine the tertiary structure of a protein.** ***(2 lines)*** (2 marks)

Total: 44 marks

Candidates' answers to Question 2

Candidate A

(a) Titrate with acid.

Candidate B

(a) Titrate with acid of known concentration.

e Candidate A scores 1 mark. Candidate B scores 2 for the extra vital detail.

Candidate A

(b) (i) First with respect to [E]; first with respect to [S]

Candidate B

(b) (i) First

e Candidate A scores both marks for a full answer. Candidate B has not written down 'first' twice, so he/she does not score the second mark.

Candidate A

(b) (ii) k[E][S]

Candidate B

(b) (ii) Rate = k[E][S]

e Candidate A has not written the complete rate equation and only scores 1 mark. Candidate B's equation is complete and scores both marks.

Candidate A

(b) (iii) 2

Candidate B

(b) (iii) Second

e Each candidate scores the mark.

Candidate A

(b) (iv) mol dm^{-3} s^{-1}

Candidate B

(b) (iv) $k = \frac{\text{rate}}{[\text{E}]\ [\text{S}]} = \frac{\text{mol dm}^{-3}\ \text{s}^{-1}}{\text{mol dm}^{-3} \times \text{mol dm}^{-3}} = \text{dm}^3\ \text{mol}^{-1}\ \text{s}^{-1}$

e Candidate A does not score as this answer is wrong and he/she has not shown any working. Candidate B is correct and scores both marks.

Candidate A

(b) (v)

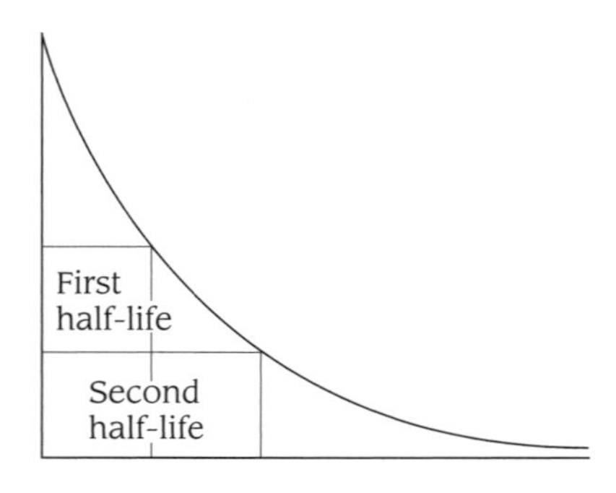

Candidate B

(b) (v)

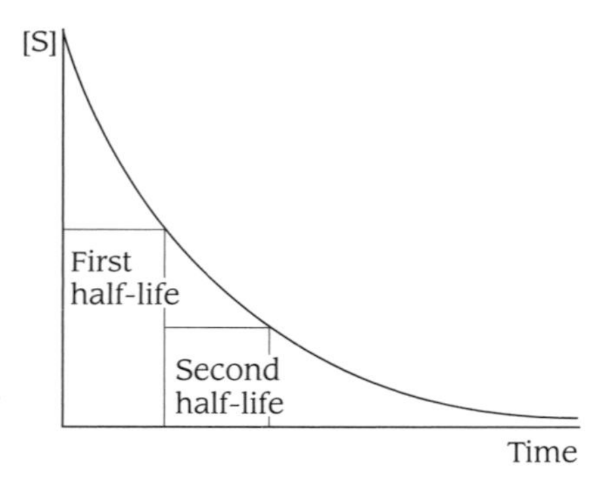

Both candidates score. Candidate A has not marked the second half-life correctly but has the correct shape for the curve. The axes are not labelled, which is risky, though in this case there are no marks available for these labels because of the way the question is worded. Candidate A scores 1 mark. Candidate B's answer is completely correct, for 2 marks.

Candidate A

(b) (vi) Zero since the rate does not change when the concentration increases.

Candidate B

(b) (vi) Zero since it does not change with concentration.

Candidate A gives the correct reason and scores the mark. Candidate B's answer is too brief. It is the rate (not the order, as implied) that does not change with time. (Avoid using 'it'.)

Candidate A

(b) (vii) The enzyme becomes saturated.

Candidate B

(b) (vii) Immediately a product molecule leaves the enzyme, another substrate molecule becomes attached, so the rate-determining step is the breakdown of the enzyme–substrate complex.

Candidate A has part of the answer and scores 1 mark, but there is no further explanation. Candidate B gives this explanation and scores both marks.

Candidate A

(c) (i) Amine

Candidate B

(c) (i) Amide

Candidate A has made a common error by confusing amines with amides. Learn functional groups *carefully*. (Here, remembering that *polyamides* exist might help.) Candidate B is correct, for 1 mark.

Candidate A

(c) (ii) $O{=}C(CH_3)(NH_2) + H_2O \rightarrow CO_2 + NH_3 + CH_4$

Candidate B

(c) (ii) $CH_3CONH_2 + H_2O \rightarrow CH_3COOH + NH_3$

Candidate A has not recalled the hydrolysis of amides and therefore invented an equation based on the hydrolysis of urea. It is not right, but it scores 1 mark for giving ammonia as one of the products. Candidate B scores both marks for the correct equation.

Candidate A

(c) (iii) The enzyme is the lock, the substrate is the key. Compound A does not fit the lock, the substrate does.

Candidate B

(c) (iii) The enzyme has an active site (a cleft in the molecule) to which the urea binds by weak intermolecular forces. CH_3CONH_2 will not fit into this site, emphasising the importance of two NH_2 groups in the binding.

Candidate A's answer is simplistic. He/she scores 2 marks for the idea of the substrate fitting and compound A not fitting and scores the 'quality of written communication' mark as the two points link together, so scores 3 marks out of 5. Candidate B gives a good answer that scores all 5 marks, including the 'quality of written communication' mark for linking all the points together. It is best not to use the 'lock and key' analogy in such answers, as it easily leads to oversimplification.

Candidate A

(c) (iv) An inhibitor binds on to the enzyme's active site.

Candidate B

(c) (iv) An inhibitor binds on to the enzyme's active site, stopping substrate molecules from getting there.

Candidate A scores 1 mark for the idea of binding to the active site. Candidate B has made the important second point that the substrate is excluded and so scores the second mark as well.

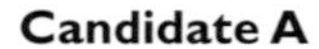

Candidate A

(c) (v) pH change breaks the hydrogen bonds that hold the tertiary structure and the active site in shape.

Candidate B

(c) (v) Heating above a certain temperature breaks the hydrogen bonds holding the tertiary structure, and hence the active site, in shape.

Candidate A scores 2 out of 3 marks by naming a correct condition and saying how denaturation works. However, changes in pH do not have much effect on hydrogen bonds — they affect ionic interactions — so he/she loses the third mark. Candidate B gives the correct cause of denaturation at high temperature so scores 3 marks.

Candidate A

(c) (vi) They use less energy.

Candidate B

(c) (vi) They often require less toxic reagents.

Both answers are correct and score the mark.

Candidate A

(d) (i) Hydrolysis.

Candidate B

(d) (i) Chromatography

Candidate A has not read the stem of the question correctly — the protein has already been hydrolysed, so this scores nothing. Candidate B gains the mark for the correct answer.

Candidate A

(d) (ii)

$$\begin{array}{ccc} & H & \\ & | & \\ {}^{+}H_2N - & C & - COO^{-} \\ & | & \\ & H & \end{array}$$

Candidate B

(d) (ii)

$$\begin{array}{ccc} & H & \\ & | & \\ {}^{+}H_3N - & C & - COO^{-} \\ & | & \\ & H & \end{array}$$

Candidate A scores 1 mark only, as he/she has not realised that NH_2 becomes NH_3^+ in forming the zwitterion. Candidate B scores both marks.

Candidate A

(d) (iii)

$$\begin{array}{ccccc} & H & & CH_3 & \\ & | & & | & \\ H_2N - & C & - C - O - HN - & C & - COOH \\ & | & & | & \\ & H & & H & \end{array}$$

Candidate B

(d) (iii)

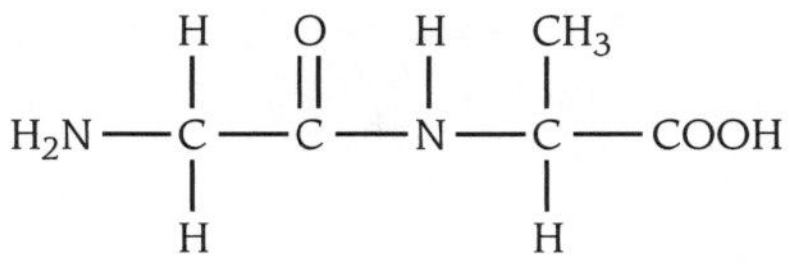

$$H_2N-\overset{\overset{H}{|}}{\underset{\underset{H}{|}}{C}}-\overset{\overset{O}{\|}}{C}-\overset{\overset{H}{|}}{N}-\overset{\overset{CH_3}{|}}{\underset{\underset{H}{|}}{C}}-COOH$$

Both candidates score 2 marks. Candidate B's representation of the peptide bond is safer, but Candidate A still scores because the question does not specify that full structural formulae are required.

Candidate A

(d) (iv) Elimination

Candidate B

(d) (iv) Condensation

Although water is eliminated, this is not the definition of an elimination reaction, so Candidate A does not score. Candidate B gains the mark.

Candidate A

(d) (v) Amide

Candidate B

(d) (v) Peptide

Both candidates are correct for 1 mark. These bonds can be called either (secondary) amides or peptides.

Candidate A

(d) (vi)

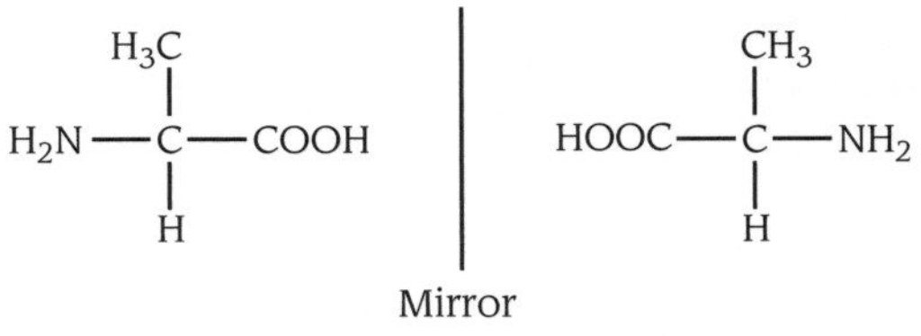

$$H_2N-\overset{\overset{H_3C}{|}}{\underset{\underset{H}{|}}{C}}-COOH \quad \Bigg| \quad HOOC-\overset{\overset{CH_3}{|}}{\underset{\underset{H}{|}}{C}}-NH_2$$

Mirror

Candidate B

(d) (vi)

H₃C above C; H_2N and COOH on wedge bonds; H on hashed bond | mirror | CH_3 above C; HOOC and NH_2 on wedge bonds; H on hashed bond

Mirror

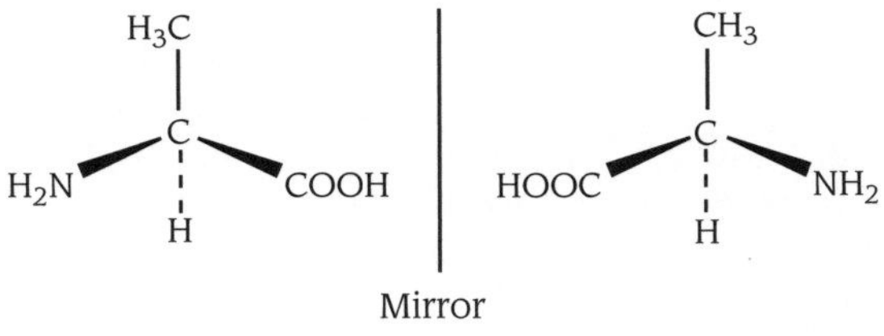

Candidate A has indicated that the isomers are related as mirror images but has not shown three-dimensional structures, so scores 1 mark. Candidate B has done this and scores 2 marks.

Candidate A

(e) (i) Primary is the chain; secondary is how the chain folds and tertiary is how the folded chain is twisted, held together by intermolecular forces between R groups.

Candidate B

(e) (i) Primary is the sequence of amino acids. Secondary is held together by hydrogen bonds. Tertiary is how the chain is twisted up.

Candidate A scores all 3 marks. His/her last phrase is not strictly necessary to answer this question. Candidate B's answer scores 1 for the primary structure but then is too vague to score further marks.

Candidate A

(e) (ii) The sequence of amino acids determines how the enzyme folds which determines the shape of the active site.

Candidate B

(e) (ii) It controls how it folds up which enables the active site to appear.

Both candidates have given sufficient information to score 2 marks.

Candidate A

(e) (iii) Hydrogen and dipole.

Candidate B

(e) (iii) Ionic interactions and covalent bonds.

Candidate A has been too brief and fails to score either mark. The interactions must be described as *hydrogen bonds* and *permanent dipole–permanent dipole bonds*. Candidate B has chosen two different interactions and named them completely, so he/she scores 2 marks.

Question 3

A biodegradable plastic

'PHB' is a natural polyester made by certain bacteria. They use it as a reserve food supply. Plastic articles made from PHB are biodegradable. The structures of PHB and its monomer are:

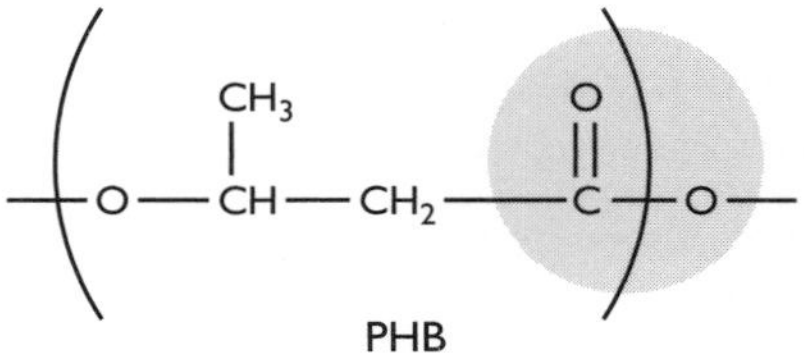

$$\left(-O-\underset{|}{\overset{CH_3}{}}CH-CH_2-\overset{O}{\overset{||}{C}}- \right)-O-$$

PHB

$$H-O-\overset{CH_3}{\overset{|}{CH}}-CH_2-\overset{O}{\overset{||}{C}}-O-H$$

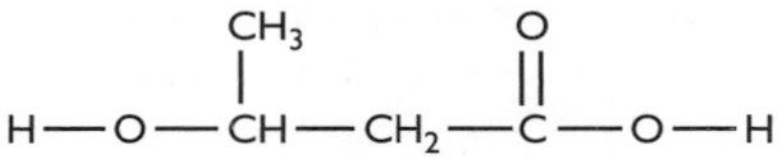

PHB monomer

(a) (i) Name the functional group that is shaded in the structure of PHB. *(1 line)* (1 mark)

(ii) Give the systematic name of the PHB monomer. *(1 line)* (2 marks)

(iii) Suggest conditions for the hydrolysis of PHB to its monomer in the laboratory. *(2 lines)* (2 marks)

(iv) What technique would be used to purify a sample of the solid monomer made in this way? *(1 line)* (1 mark)

(v) Suggest conditions for the synthesis of PHB from its monomer in the laboratory. *(2 lines)* (2 marks)

(vi) Draw the structural formula of the salt formed when the PHB monomer reacts with sodium hydroxide. *(space)* (2 marks)

(b) Nylon Y could be made from the monomer X shown below.

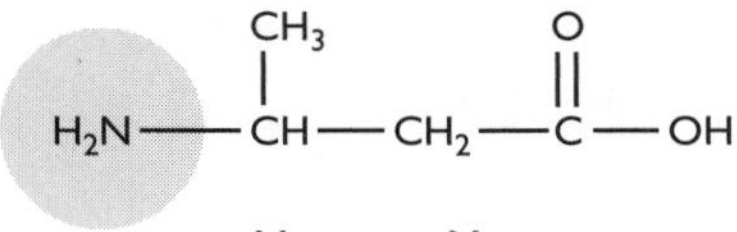

$$H_2N-\overset{CH_3}{\overset{|}{CH}}-CH_2-\overset{O}{\overset{||}{C}}-OH$$

Monomer X

(i) Name the functional group shaded in monomer X. *(1 line)* (1 mark)

(ii) Name the compound $C_2H_5NH_2$ that contains this functional group. *(1 line)* (1 mark)

(iii) Draw the structure of the repeating unit of nylon Y. *(space)* (2 marks)

(c) Another nylon is made by reacting the two monomers shown.

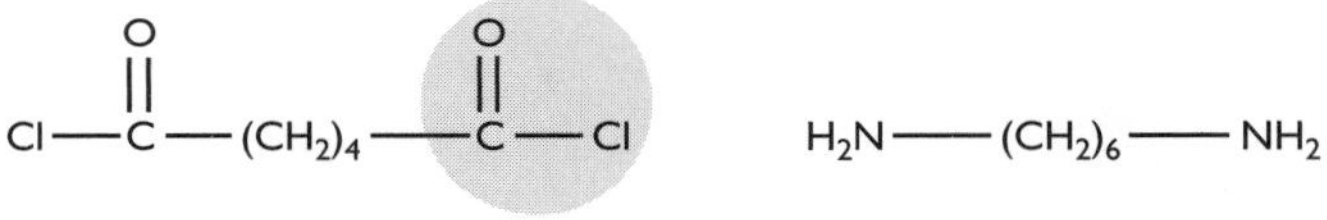

$$Cl-\overset{O}{\overset{||}{C}}-(CH_2)_4-\overset{O}{\overset{||}{C}}-Cl \qquad H_2N-(CH_2)_6-NH_2$$

(i) Name the functional group that is shaded in the left-hand structure. ***(1 line)*** (1 mark)

(ii) Why is this functional group rather than a carboxylic acid group often used for laboratory synthesis of nylon? ***(2 lines)*** (1 mark)

(iii) Name the small molecule that is produced when the two monomers react. ***(1 line)*** (1 mark)

(d) This part of the question compares the properties of PHB with those of nylon Y.

(i) Name the strongest type of intermolecular force found between the chains of:

- **PHB** ***(1 line)***
- **Nylon Y** ***(1 line)*** (2 marks)

(ii) A sample of nylon Y and a sample of PHB of the same shape and chain length are prepared. Which would you expect to:

- **be less flexible?**
- **have the higher T_g?**

Give reasons for your answers. ***In your answers you should link your explanations to the properties concerned. (5 lines)*** (4 marks)

(iii) A sample of nylon can be ***cold drawn.*** **Describe how this is done and say how it will change the structure and properties of nylon.** ***(3 lines)*** (2 marks)

(e) Explain why biodegradable polymers are desirable. ***(2 lines)*** (1 mark)

(f) The ability to make PHB is coded in the DNA of a bacterial cell nucleus. These genes can be placed in the nuclei of cells of plants such as cotton, which will then make PHB.

(i) Using symbols labelled 'base', 'sugar' and 'phosphate', draw the structure of the ***double chain*** **of DNA. Show where the chains are linked.** ***(space)*** (3 marks)

(ii) Name the intermolecular bonds that link the two chains. ***(1 line)*** (1 mark)

(iii) Use your ***Data Sheets*** **to show a molecule of cytosine linked to a molecule of deoxyribose.** ***(space)*** (2 marks)

(iv) Which part of the DNA structure carries the code for a single amino acid in a protein structure? ***(2 lines)*** (2 marks)

(v) Give ***one*** **advantage of a country maintaining a comprehensive DNA database of its citizens.** ***(1 line)*** (1 mark)

Total: 35 marks

Candidates' answers to Question 3

Candidate A

(a) (i) Carbonyl

Candidate B

(a) (i) Ester

Candidate A has not looked at the whole group. C=O is carbonyl on its own but here it is part of an ester group. Candidate A does not score but Candidate B does.

Candidate A

(a) (ii) 2-hydroxybutanoic acid

Candidate B

(a) (ii) 3-hydroxybutanoic acid

Candidate A is almost correct and scores 1 mark for 'hydroxybutanoic acid'. Candidate B has remembered that numbering in a carboxylic acid starts from the COOH carbon and so he/she scores both marks.

Candidate A

(a) (iii) Reflux with concentrated sulfuric acid.

Candidate B

(a) (iii) Reflux with concentrated hydrochloric acid.

Candidate A's conditions are harsh and the polymer would probably be dehydrated and turn brown. He/she scores 1 mark for heating. Candidate B's concentrated hydrochloric acid is also rather harsh but acceptable, so he/she gains 2 marks. Moderately concentrated hydrochloric acid is the best reagent.

Candidate A

(a) (iv) Melting point

Candidate B

(a) (iv) Recrystallisation

Candidate B is correct, for 1 mark. Taking a melting point is a test for purity but it does not purify a solid, so candidate A does not score.

Candidate A

(a) (v) Heat with methanol and conc. sulfuric acid.

Candidate B

(a) (v) Heat with conc. sulfuric acid.

Methanol is not required. Candidate A has not realised that the monomer has both the alcohol and the acid groups required to react with itself. Just concentrated sulfuric acid and heat are needed. This scores 1 mark only (for heating), since the mention of methanol contradicts the sulfuric acid mark. Candidate B scores 2 marks.

Candidate A

(a) (vi)

$$H{-}O{-}\underset{}{\overset{\overset{\displaystyle CH_3}{|}}{CH}}{-}CH_2{-}\overset{\overset{\displaystyle O}{||}}{C}{-}O{-}Na$$

Candidate B

(a) (vi)

$$H{-}O{-}\overset{\overset{\displaystyle CH_3}{|}}{CH}{-}CH_2{-}\overset{\overset{\displaystyle O}{||}}{C}{-}O^-$$

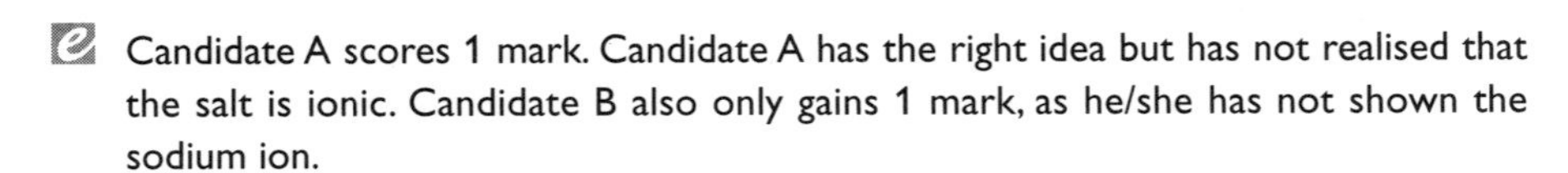

Candidate A scores 1 mark. Candidate A has the right idea but has not realised that the salt is ionic. Candidate B also only gains 1 mark, as he/she has not shown the sodium ion.

Candidate A
(b) (i) Amine

Candidate B
(b) (i) Primary amine

Amine is acceptable, primary amine gives more detail. Both score 1 mark.

Candidate A
(b) (ii) 1-aminoethane

Candidate B
(b) (ii) Aminoethane

The '1' is not necessary but it would not lose the mark. Both score.

Candidate A
(b) (iii)

$$-\underset{\underset{}{}}{\overset{H}{\overset{|}{N}}}-\overset{CH_3}{\overset{|}{CH}}-CH_2-\overset{O}{\overset{\|}{C}}-$$

Candidate B
(b) (iii)

$$H-\overset{H}{\overset{|}{N}}-\overset{CH_3}{\overset{|}{CH}}-CH_2-\overset{O}{\overset{\|}{C}}-\overset{H}{\overset{|}{N}}-\overset{CH_3}{\overset{|}{CH}}-CH_2-\overset{O}{\overset{\|}{C}}-OH$$

Candidate B has the right idea and scores 1 mark. However, he/she has not answered the question — just given the result of condensing two monomers together, rather than giving the repeating unit when several combine together to form the polymer. Candidate A scores both marks.

Candidate A
(c) (i) Ethanoyl chloride

Candidate B
(c) (i) Acyl chloride

Candidate A has given the name of one specific acyl chloride and does not score. Note how important it is to learn the names of functional groups in this unit. Candidate B is correct, for 1 mark.

Candidate A
(c) (ii) It reacts much faster.

Candidate B
(c) (ii) It is cheaper.

Candidate B has gone for a common answer. Even if it were correct, it should be qualified with a reason. However, this is wrong, as chlorine-containing compounds are more expensive. Candidate A is correct, for 1 mark.

Candidate A

(c) (iii) Hydrogen chloride

Candidate B

(c) (iii) Hydrochloric acid

Candidate A is correct and scores the mark. Candidate B misses the mark by a whisker. Hydrochloric acid is not a small molecule but a solution containing ions.

Candidate A

(d) (i) Hydrogen bonding; more hydrogen bonding

Candidate B

(d) (i) pd–pd forces; hydrogen bonds

Candidate A scores 1 mark for naming the strongest intermolecular bond in nylon. In PHB, the bonds are permanent dipole–permanent dipole, since there are no –OH groups. However, Candidate B still only scores 1 mark as abbreviations are not permitted.

Candidate A

(d) (ii) Nylon would be less flexible and have the higher T_g because it has more hydrogen bonds.

Candidate B

(d) (ii) Nylon has the stronger intermolecular forces, so it is less flexible. It also means that more energy is needed to separate the chains so its T_g will be higher.

Candidate A should realise that a short answer like this is unlikely to gain 4 marks. He/she scores 1 mark for identifying nylon as less flexible and having the higher T_g and 1 mark for explaining this in terms of intermolecular bonds. Candidate B scores 3 marks. He/she has left out the important step of explaining low flexibility — *the chains cannot move over each other easily.* This is the crucial 'link' point that is needed to score the 'quality of written communication' mark.

Candidate A

(d) (iii) Nylon is slowly pulled out. This makes the structure more crystalline and thus stronger as the chains lie closer together.

Candidate B

(d) (iii) Nylon is pulled out and becomes stronger.

Candidate A scores 3 marks for a good answer. Candidate B scores 2, but has left out the point about the change to the structure of nylon.

Candidate A

(e) They can be broken down when buried in the ground.

Candidate B

(e) They can be broken down quickly by bacteria and thus avoid causing the pollution caused by most plastics which take years to break down.

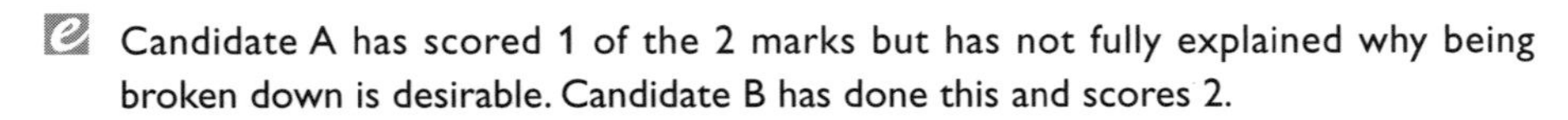

Candidate A has scored 1 of the 2 marks but has not fully explained why being broken down is desirable. Candidate B has done this and scores 2.

Candidate A

(f) (i)

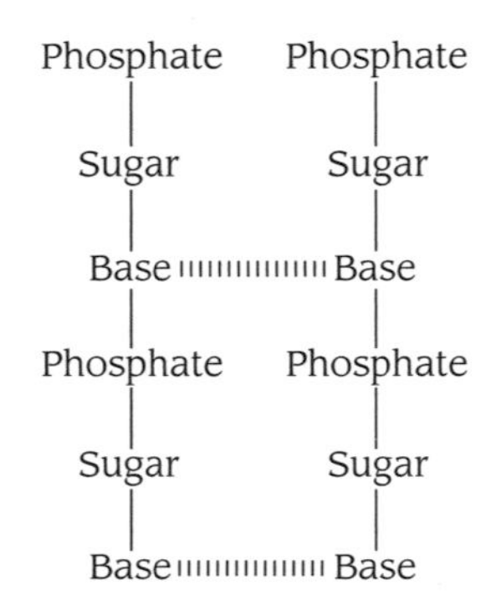

Candidate B

(f) (i)

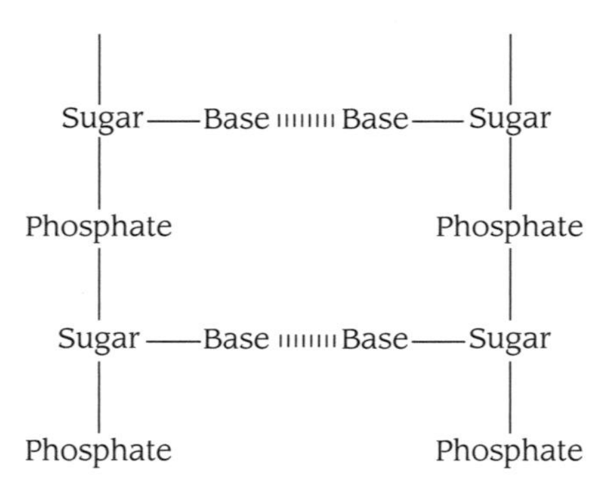

Candidate A is confused. Candidate A understands that the two chains are linked through the bases and so scores 1 mark for this, but has lost the other two marks by putting the bases in the chain, instead of sticking out from the sugars. Candidate B scores 3 marks for having everything in the right place.

Candidate A

(f) (ii) Hydrogen bonds

Candidate B

(f) (ii) Weak intermolecular bonds

Candidate A scores the mark, while candidate B is too vague and does not.

Candidate A

(f) (iii)

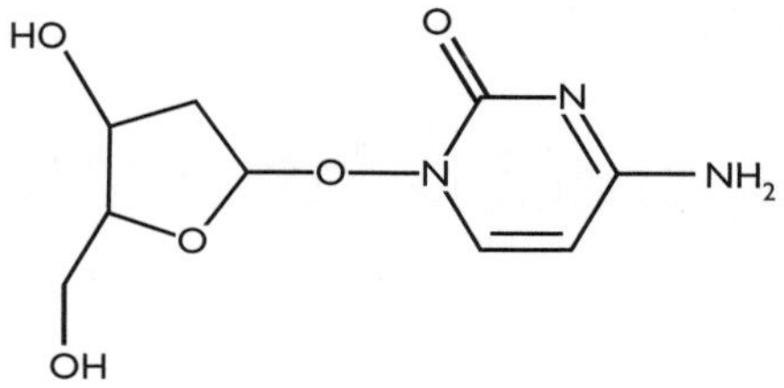

HO ... O—N ... N ... NH_2 ... O ... OH

Candidate B

(f) (iii)

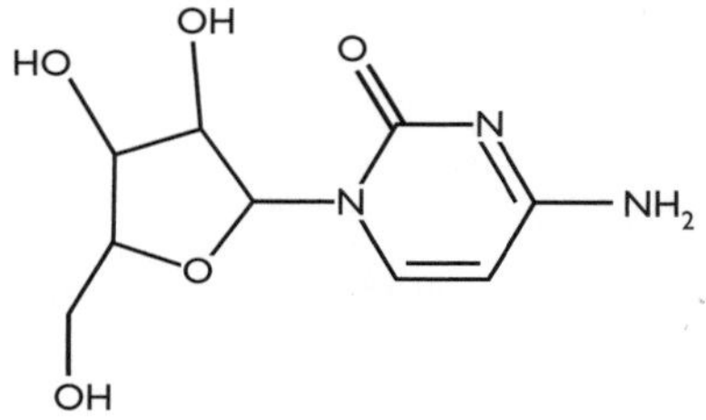

OH ... HO ... O ... N ... N ... NH_2 ... O ... OH

Both candidates score 1 out of 2. Candidate A has the correct molecules, combined in the right places but has not realised that the condensation reaction by which they combine takes the OH from the sugar and the H from the NH on the base. Candidate B has this correct but has made a careless error and used the structure of ribose not deoxyribose.

Candidate A

(f) (iv) The bases

Candidate B

(f) (iv) Three adjacent bases

Candidate A scores 1 for knowing it is something to do with bases. Candidate B scores the second mark for knowing that the code is three bases.

Candidate A

(f) (v) Many more crimes can be solved.

Candidate B

(f) (v) Only those convicted of crimes should have their data stored.

Again, candidate B has made a careless mistake and not read the question carefully. The statement may be true but it does not answer the question, so it does not score. Candidate A has given a good answer that does answer the question, so scores 1 mark.

Question 4

Preventing steel boats from rusting

Many steel boats have blocks of zinc fixed to their hulls to prevent them rusting. The blocks corrode rather than the iron. Some relevant standard electrode potentials are:

		$E^{\ominus}$/V
A	$Zn^{2+}(s) + 2e^- \rightarrow Zn(aq)$	−0.76
B	$Fe^{2+}(aq) + 2e^- \rightarrow Fe(s)$	−0.44
C	$Sn^{2+}(aq) + 2e^- \rightarrow Sn(s)$	−0.14
D	$2H^+(aq) + 2e^- \rightarrow H_2(g)$	0 (by definition)
E	$O_2(g) + 2H_2O(l) + 4e^- \rightarrow 4OH^-(aq)$	+0.40

(a) A cell is set up using half-equations A and C. Draw a labelled diagram of this cell under standard conditions. ***(space)*** (4 marks)

(b) The two half-equations involved in rusting are B and E.

(i) Explain how these two electrode potentials and their position in the table show that iron will react with oxygen and water. ***(3 lines)*** (3 marks)

(ii) Write the equation for the reaction that occurs in part (i). ***(space)*** (2 marks)

(iii) Calculate the value of $E^{\ominus}_{cell}$ for this reaction. ***(space)*** (1 mark)

(c) The diagram below shows a zinc block fixed to a steel hull immersed in seawater.

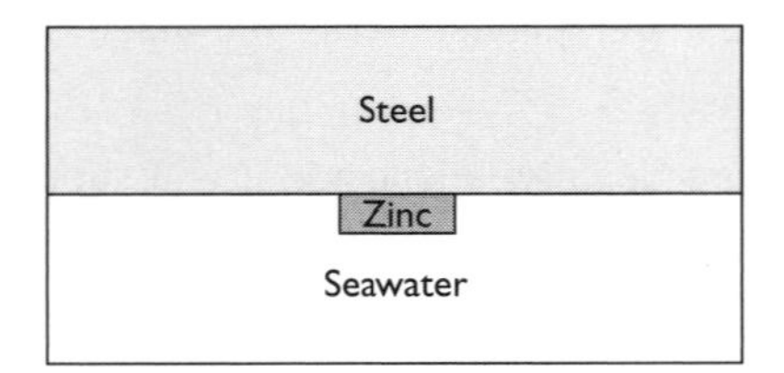

Consider what is happening using the electrode potentials above.

(i) Mark the flow of electrons on the diagram. (1 mark)

(ii) Explain why this inhibits the rusting of the iron. ***(3 lines)*** (2 marks)

(d) Describe and explain another method of protecting steel from rusting that does not involve using a reactive metal. Explain one disadvantage of this method. ***(3 lines)*** (3 marks)

Total: 16 marks

Candidates' answers to Question 4

Candidate A

(a)

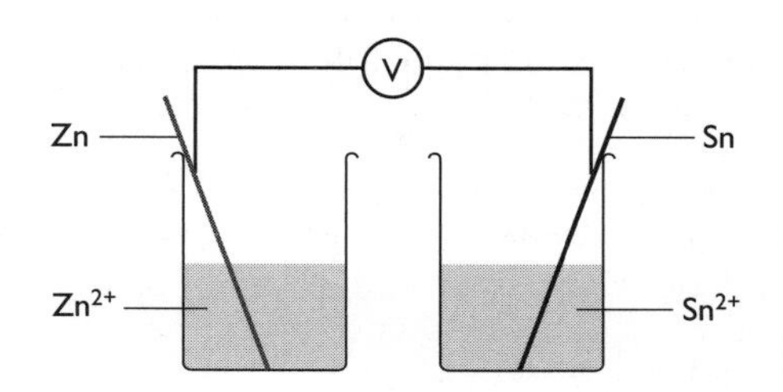

Candidate B

(a)

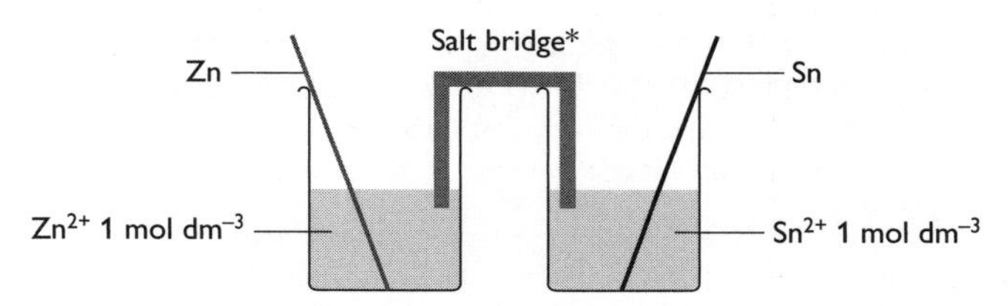

*Filter paper plus saturated potassium nitrate solution

Candidate A has left out the salt bridge and the concentrations of the reagents, so he/she scores 2 out of 4 marks. Candidate B has left out the voltmeter, but this does not matter, because the question asks just for a diagram of the cell. Candidate B scores all 4 marks.

Candidate A

(b) (i) Since B is above E it will go backwards and make E go forwards.

Candidate B

(b) (i) Since B is the more negative electrode it will supply electrons through the external circuit to E. Thus the reaction in B will be reversed to supply electrons and the reaction in E will go as printed.

Candidate A's explanation is incomplete. However, candidate A has made the point about reaction B being reversed and E going in the right direction, and scores 1 mark. Candidate B has done much better. The crucial part of the argument is that *electrons flow from the more negative electrode (in this case the iron) to the more positive one through the external circuit.* Candidate B has included this and clearly stated which is the more positive electrode, and scores 3 marks.

Candidate A

(b) (ii) $Fe(s) + O_2(g) + 2H_2O(l) \rightarrow 4OH^-(aq) + Fe^{2+}(aq)$

Candidate B

(b) (ii) $2Fe + O_2 + 2H_2O \rightarrow 4OH^- + 2Fe^{2+}$

Candidate A has the right idea but has not balanced the equation. Candidate B scores both marks. He/she has left out the state symbols, but they were not asked for here.

Candidate A

(b) (iii) $E_{cell} = 0.40 - 0.44 = -0.04$ V

Candidate B

(b) (iii) $E_{cell} = 0.4 - (-0.44) = 0.84$

Candidate A has failed to realise that the electrode potentials are on either side of zero, thus their difference is 0.84, not 0.04. However, he/she scores 1 mark for the correct units. Candidate B has the correct value for $E^{\ominus}_{cell}$ (the sign does not matter here) but has failed to give the units, so he/she loses the second mark.

Candidate A

(c) (i)

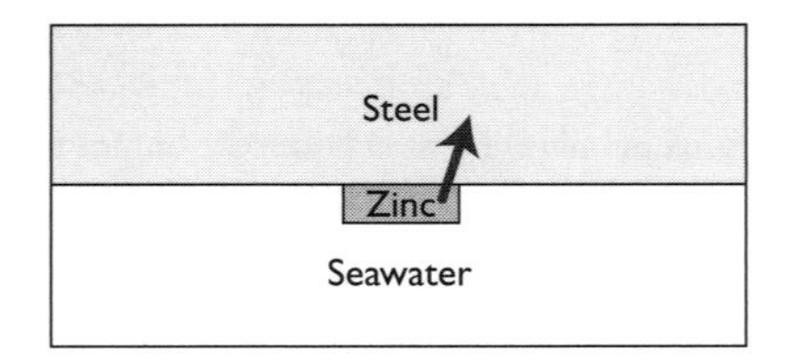

Candidate B

(c) (i)

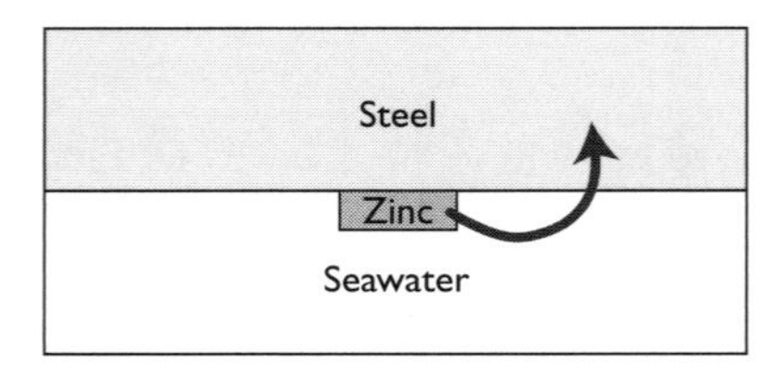

Candidate A is correct, for 1 mark. Electrons only flow through metals; they never flow through solutions (ions move through solutions). Candidate B does not score.

Candidate A

(c) (ii) This inhibits rusting because the iron wants to lose electrons and in this situation they flow towards it.

Candidate B

(c) (ii) Electrons are forced onto the iron by the zinc, which corrodes instead.

e Candidate A has given a good answer but it is not quite enough for full marks. The idea of zinc corroding is needed for the second mark. Candidate B scores both marks.

Candidate A

(d) Covering an object with grease stops it rusting. It stops working if the layer is broken, in fact it rusts faster then, like at the bottom of a pit.

Candidate B

(d) Steel can be covered with a polymer coat to stop it rusting. This prevents the access of the water or oxygen. It is expensive, however.

e Candidate A scores 2 marks because he or she has *described* the method but not *explained* it, so missing the second mark for the explanation. He/she has then explained the disadvantage well and scores the third mark. Candidate B scores the first 2 marks but his/her attempt at the third is weak and does not score.

Question 5

Liniment

Liniment used by athletes often contains 'oil of wintergreen', methyl salicylate.

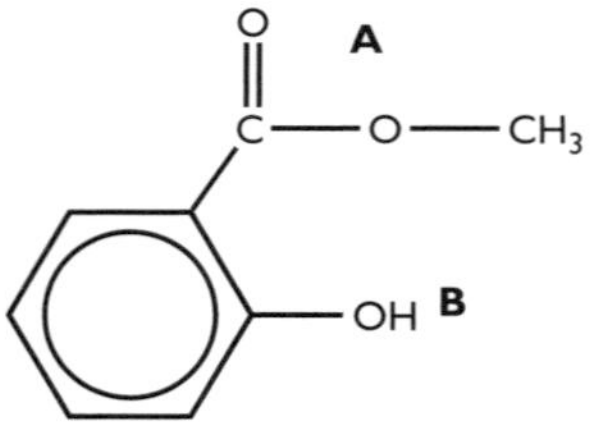

(a) Identify the functional groups in methyl salicylate, which are labelled A and B. ***(2 lines)*** (1 mark)

(b) Describe a test and its result that would identify functional group B. ***(3 lines)*** (2 marks)

(c) A chemist has available the following compounds:

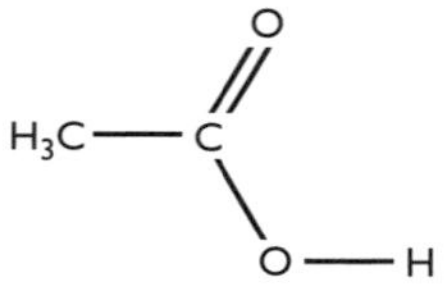

C: $H_3C-COOH$

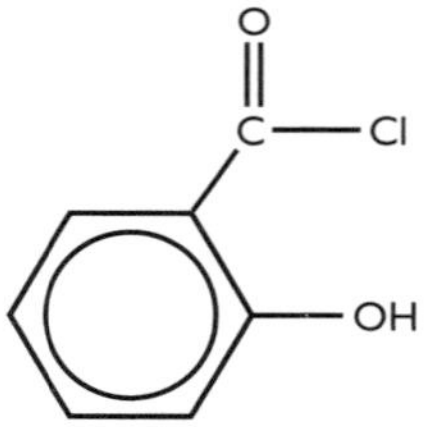

D: 2-hydroxybenzoyl chloride (benzene ring with $-COCl$ and $-OH$)

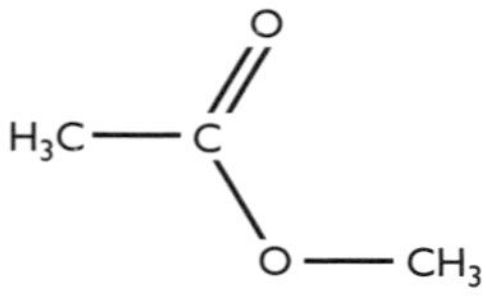

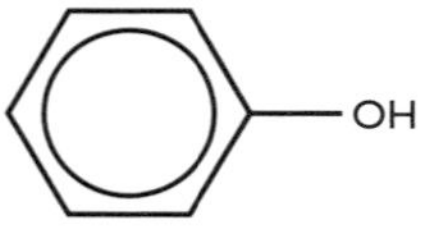

E: CH_3OH

F: $H_3C-COO-CH_3$

G: phenol (benzene ring with $-OH$)

(i) **Name compounds C and F.** ***(2 lines)*** (2 marks)

(ii) **Which two of these compounds could the chemist react together to make methyl salicylate?** ***(1 line)*** (1 mark)

(iii) **Compound C is acidic in solution and reacts with sodium carbonate solution. Write equations to illustrate the acidity and the reaction with carbonate.** ***(2 lines)*** (2 marks)

(iv) **Draw the structure of the compound formed by reacting compound D with compound G.** ***(space)*** (2 marks)

(d) Another chemist boiled some methyl salicylate under reflux with moderately concentrated hydrochloric acid.

(i) Draw a labelled diagram to show how you would heat the mixture under reflux. *(space)* (2 marks)

(ii) Describe how thin layer chromatography could be used to discover whether any starting compound remained at the end of the refluxing period. *(space)* (4 marks)

(iii) The products of the reaction are separated. One has an M_r of 138. Draw the structure of this compound and name the spectroscopic technique that would be used to determine the M_r. *(space, line)* (3 marks)

(iv) One of the products gives the infrared spectrum shown below. Identify this product, giving your reasons. *(5 lines)* (3 marks)

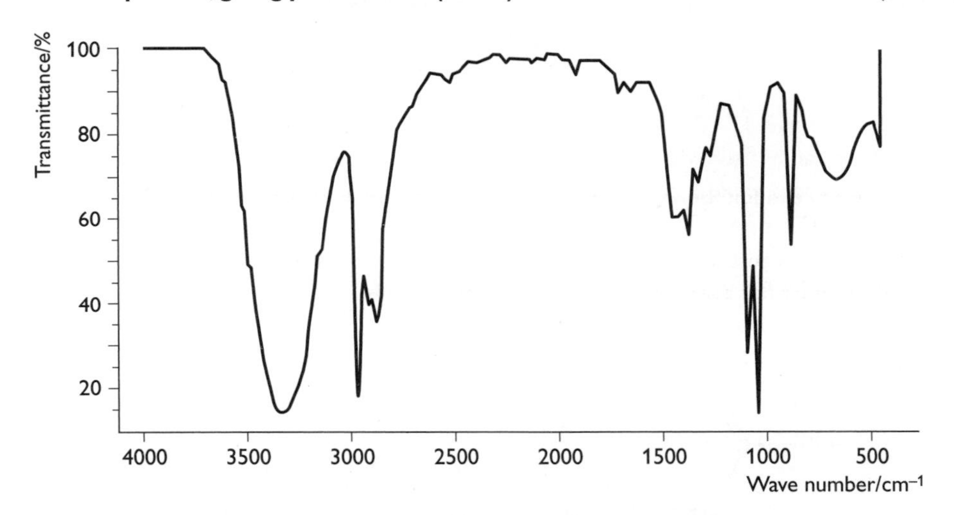

(e) If methyl salicylate had recently been discovered, clinical trials would be carried out as a stage of testing it. Give three questions these trials would ask. *(3 lines)* (3 marks)

(f) Many medicines can be improved by modifying their structures. Suggest how combinatorial chemistry could be used to make a set of compounds similar to methyl salicylate. *(2 lines)* (1 mark)

Total: 26 marks

Candidates' answers to Question 5

Candidate A

(a) A is a methyl ester. B is an alcohol.

Candidate B

(a) A is an ester. B is a phenol.

e Both candidates score the mark for group A, though 'ester' is sufficient. Only candidate B scores the mark for group B, as a hydroxy group attached to a benzene ring is called a phenol, not an alcohol.

Candidate A

(b) Slightly acidic in solution

Candidate B

(b) Reacts with neutral iron(III) chloride to give a purple colour.

Candidate A scores 1 mark out of 2. Candidate A has not said how to carry out the test. Candidate B has given the best test for a phenol and described it fully, so he/she scores 2 marks.

Candidate A

(c) (i) C is ethanoic acid; F is methyl ethanoate.

Candidate B

(c) (i) C is ethanoic acid; F is ethyl methanoate.

Both candidates have identified compound C correctly and score 1 mark. Candidate A has compound F correct and scores another mark. Candidate B has confused the parts of the ester. The $CH_3C(=O)—$ group comes from ethanoic acid and the OCH_3 comes from methanol — hence methyl ethanoate.

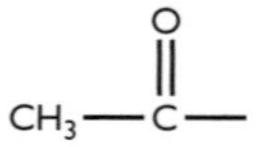

Candidate A

(c) (ii) Compounds F and G.

Candidate B

(c) (ii) Compounds D and E.

Candidate B is correct and scores the mark. To make methyl salicylate, the acyl chloride group in D reacts with the alcohol E to form an ester. Candidate A scores nothing as he or she has simply selected two likely looking molecules without considering the chemical reaction.

Candidate A

(c) (iii) $CH_3COOH \rightarrow CH_3COO^- + H^+$

$2CH_3COOH + Na_2CO_3 \rightarrow 2CH_3COONa + CO_2 + H_2O$

Candidate B

(c) (iii) $CH_3COOH \rightarrow CH_3COO^- + H^+$

$2CH_3COOH + CO_3^{2-} \rightarrow 2CH_3COO^- + CO_2 + H_2O$

Both candidates score 2 marks. They both have the first equation correct and give two alternative correct answers for the second.

Candidate A

(c) (iv)

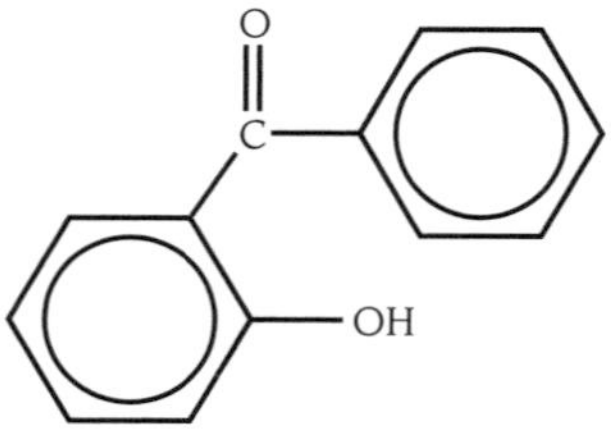

Candidate B
(c) (iv)

Candidate A scores 1 mark for identifying the correct place at which the reaction occurs. However, he/she has not shown an ester being formed. Candidate B has done this and scores 2 marks.

Candidate A
(d) (i)

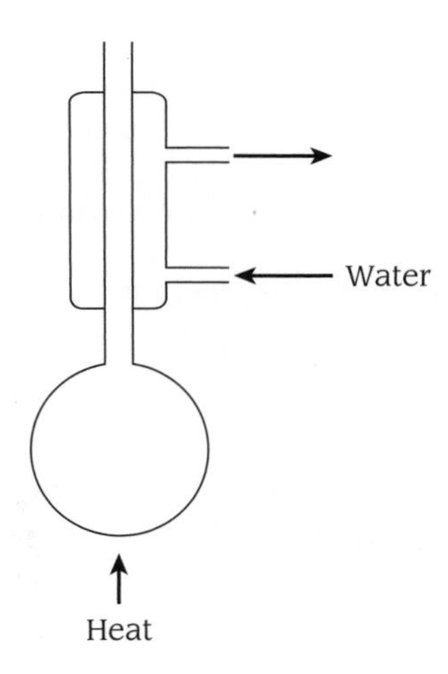

Candidate B
(d) (i)

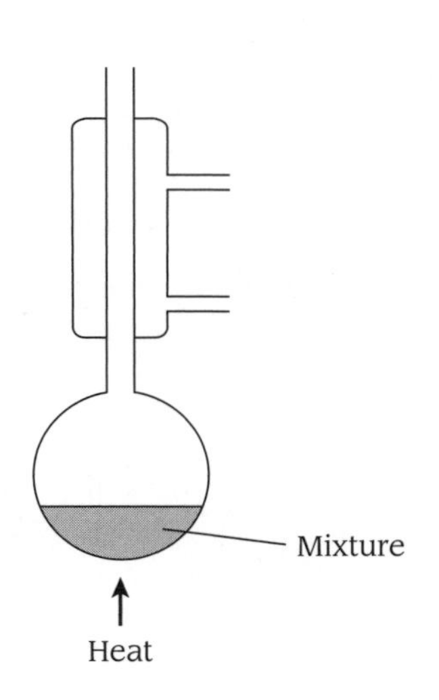

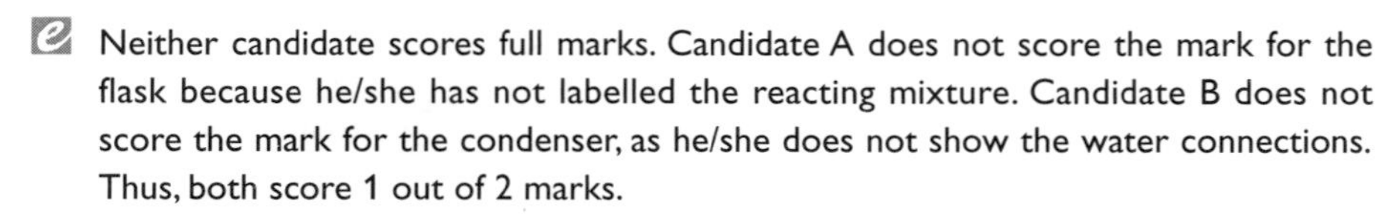

Neither candidate scores full marks. Candidate A does not score the mark for the flask because he/she has not labelled the reacting mixture. Candidate B does not score the mark for the condenser, as he/she does not show the water connections. Thus, both score 1 out of 2 marks.

Candidate A

(d) (ii) Spot, onto a thin layer plate, some mixture after reflux and some methyl salicylate. Run the chromatogram. If a dot appears at the same height as the methyl salicylate dot, this means that some remains.

Candidate B

(d) (ii) Set up the apparatus as shown in the diagram.

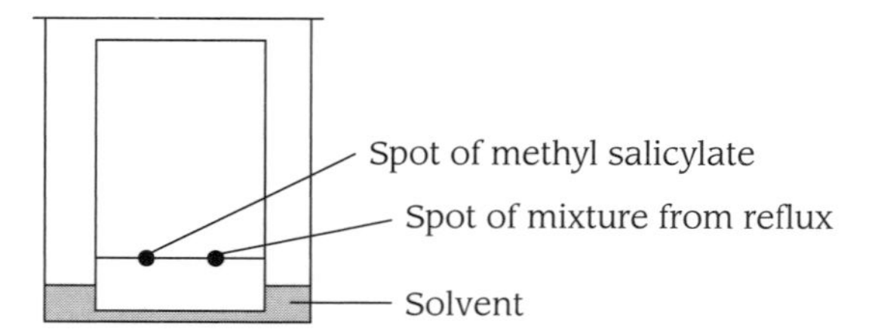

The solvent is allowed to rise up the plate. The plate is then dried and viewed under UV light. If a spot is seen above the reflux mixture that is on a level with the methyl salicylate spot, then some unreacted compound remains.

It is always best to draw a diagram and thus Candidate A has missed out on some easy marks for the method, such as having the solvent level below the spots and covering the beaker. He/she has also failed to say how the spots would be located, but scores 2 marks for the correct spots and the correct interpretation. Candidate B has given a full answer (with a diagram, which makes some of the points not described in writing), so he/she scores all 4 marks.

Candidate A

(d) (iii) RMM of methyl salicylate is 154. Loss of 16 gives 138, so this must be an oxygen atom.

Mass spectroscopy is used.

Candidate B

(d) (iii) Methyl salicylate is hydrolysed to salicylic acid and methanol. The M_r of methanol is much smaller than 138, so the compound must be salicylic acid. Mass spectroscopy is used.

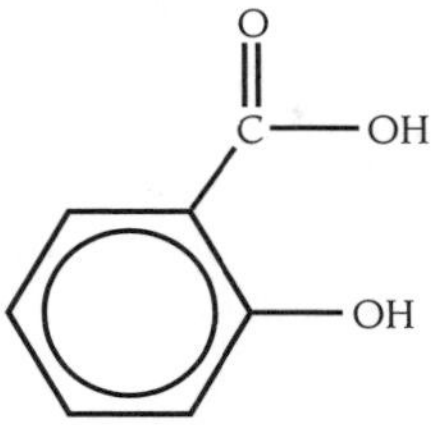

e Candidate A has gone off on the wrong track by miscalculating the M_r of methyl salicylate (it is 152 — could this be due to a misunderstanding of the number of hydrogen atoms in the substituted benzene ring — four not six?). Candidate A has also looked for fragments from mass spectrometry, which the question does not require. He/she does not score any marks for this part, but scores one for naming mass spectroscopy as the technique. Candidate B has given the correct answer, though he/she is a bit cavalier in the reasoning — it would have been safer to check that the M_r of salicylic acid was 138. Candidate B scores all 3 marks.

Candidate A

(d) (iv) This compound has O–H groups (3400 cm^{-1}) and C–H groups (2900 cm^{-1}). Thus it could be either an alcohol or an acid.

Candidate B

(d) (iv) The compound has O–H groups but no C=O groups. Thus it is methanol.

e Candidate A scores 2 marks for identifying two bonds present. However, he/she does not score anything for the deduction. The compound could not be an acid because of the lack of C=O. Candidate B scores the 2 marks for the functional groups, though it would be safer to give the values of the absorbances. Candidate B also goes on to identify the compound correctly from his/her understanding of the hydrolysis of esters, thus scoring the third mark too.

Candidate A

(e) Is it safe and does it work better than the standard treatment?

Candidate B

(e) Is it safe? Does it work? Is it better than the standard treatment?

e Both score 3 marks. Candidate A has dangerously elided two of the questions but gets away with it here.

Candidate A

(f) Make lots of compounds very quickly and test them.

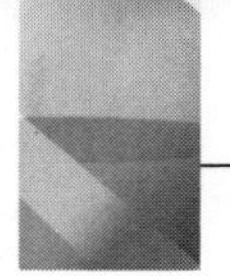

Candidate B

(f) Make lots of other esters in an automated rig and assess them by large-scale screening.

e Candidate A does not score as he/she has not mentioned what the compounds might be. Candidate B has made a sensible suggestion and gains the mark.